D0162388

About Island Press

Since 1984, the nonprofit organization Island Press has been stimulating, shaping, and communicating ideas that are essential for solving environmental problems worldwide. With more than 1,000 titles in print and some 30 new releases each year, we are the nation's leading publisher on environmental issues. We identify innovative thinkers and emerging trends in the environmental field. We work with world-renowned experts and authors to develop cross-disciplinary solutions to environmental challenges.

Island Press designs and executes educational campaigns in conjunction with our authors to communicate their critical messages in print, in person, and online using the latest technologies, innovative programs, and the media. Our goal is to reach targeted audiences—scientists, policymakers, environmental advocates, urban planners, the media, and concerned citizens—with information that can be used to create the framework for long-term ecological health and human well-being.

Island Press gratefully acknowledges major support of our work by The Agua Fund, The Andrew W. Mellon Foundation, The Bobolink Foundation, The Curtis and Edith Munson Foundation, Forrest C. and Frances H. Lattner Foundation, The JPB Foundation, The Kresge Foundation, The Oram Foundation, Inc., The Overbrook Foundation, The S.D. Bechtel, Jr. Foundation, The Summit Charitable Foundation, Inc., and many other generous supporters.

The opinions expressed in this book are those of the author(s) and do not necessarily reflect the views of our supporters.

How to Feed the World

EDITED BY
Jessica Eise and Ken Foster

Library of Congress Control Number: 2017950979

All Island Press books are printed on environmentally responsible materials.

Manufactured in the United States of America
10 9 8 7 6 5 4 3 2

Keywords: agriculture, crops, farming, fertilizers, food waste, international trade, irrigation, pesticides, soil, sustainability

CONTENTS

Introduction 1
Jessica Eise and Ken Foster

Chapter 1 Inhabitants of Earth 5
Brigitte S. Waldorf

Chapter 2 The Green, Blue, and Gray Water Rainbow 24
Laura C. Bowling and Keith A. Cherkauer

Chapter 3 The Land That Shapes and Sustains Us 46
Otto Doering and Ann Sorensen

Chapter 4. Our Changing Climate 59
Jeff Dukes and Thomas W. Hertel

Chapter 5 The Technology Ticket 77
Uris Baldos

Chapter 6 Systems 94
Michael Gunderson, Ariana Torres, Michael Boehlje, and Rhonda Phillips

Chapter 7 Tangled Trade 115
Thomas W. Hertel

Chapter 8 Spoiled, Rotten, and Left Behind 132
Ken Foster

Chapter 9 Tipping the Scales on Health 148
Steven Y. Wu

Chapter 10 Social License to Operate 165
Nicole J. Olynk Widmar

Chapter 11 The Information Hinge 176
Jessica Eise

Chapter 12 Achieving Equal Access 189
Gerald Shively

Conclusion 207
Jessica Eise and Ken Foster

Afterword 217
Acknowledgments 219
Notes 221
Contributors 235
Index 241

Introduction

Jessica Eise and Ken Foster

In the chilly Indiana winter of early 2016, we sat down for a meeting. We were in the agricultural economics building at Purdue, which has served as Indiana's land grant university since 1869, where we work with some of the world's foremost experts on food, health, and the environment. All those present at this meeting, in their individual research, were seeking answers to the enormous challenge of feeding the world sustainably.

Our meeting was an unusual one. We were seeking to determine how we could bring our core expertise together in a way that was accessible for people outside the walls of academia. We wanted people far beyond the labs, classrooms, and fields of Purdue to see how issues as varied as irrigation, tariffs, soil health, and diet decisions interconnect. In doing so, we hoped to highlight the critical challenges we must overcome to feed the world, all the while showing that a healthy, plentiful food system is possible when we take an integrated approach.

From that meeting, the idea of this book was born. Our first step, we realized, was identifying the challenges themselves. With so many factors affecting our food system, what issues were primary? Months passed after that initial meeting as we bounced ideas back and forth, consulted with colleagues, sat in on lectures, read, researched, and argued. In the end, we agreed on 12 topics:

1. Population growth
2. Water scarcity
3. Land use
4. Climate change
5. Technology adoption
6. Competing food systems
7. International trade
8. Food waste and loss
9. Health
10. Social license to operate
11. Communication
12. Achieving equal access to food

By the time we nailed these down, winter was long gone. It had then become a matter of getting the right researchers onboard. More digging, meetings, conversations, and emails ensued, from which an extraordinary team was born. To name just a few, from the Purdue Center for Climate Change Research, Jeff Dukes, who holds the Belcher Chair for Environmental Sustainability, came aboard. In the Agronomy Department, hydrologist Laura Bowling joined the team. In Agricultural and Biological Engineering, Keith Cherkauer, an environmental engineer, agreed to participate. From Horticulture, we found Ariana Torres. And within our own department, Agricultural Economics, there was a wealth of expertise, such as development expert Gerald Shively and population guru Brigitte Waldorf, and, within the Center for Global Trade Analysis, both Tom Hertel and Uris Baldos decided to contribute.

One by one, we built the team until we reached a total of 17 participants. Together, we had managed to form a group comprising some of the most cutting-edge researchers in the field of food and agriculture. The group, in terms of both expertise and personal backgrounds, is tremendously diverse. Ranging from Germany, the Philippines, and Ecuador to the United States and beyond, their origins and life experiences shaped their interests in the subject matter they address. In many cases, they weave their own personal experiences into their chapters, connecting the big issues to our tangible, day-to-day existence. With knowledge spanning concepts as tiny as the water molecule to as great as Earth's atmosphere, our collective range of expertise grew to be as vast as the planet itself.

Holding together a group of 17 scientists and researchers, while they all juggle so many competing demands for their time, proved to be an undertaking in itself. A stubborn focus on the end goal was the glue that held everyone together the following year of writing, deadlines, and rounds of revisions. That end goal was more than just a book. It was the goal of sharing information beyond the confines of the Ivory Tower, motivated by a belief in the importance of knowledge that is accessible for everyone. We live in a society today where the boundaries between fact, fiction, and opinion are easily blurred, yet food security is far too tenuous, and far too important, to depend on shadowy misconceptions.

This book is an objective resource on what we really face in food and agriculture. It lays out our greatest challenges in a clear, straightforward way that is available for those who are interested. The issues herein are straight from the mouths of scientists and researchers. We didn't pay anyone to contribute to this book. Nobody's research was funded as a result. Every contributor played a part because collectively they understand that our future depends on our ability to work together empowered by knowledge—and that members of our society deserve a trusted resource. With little incentive to participate, their contributions to this daunting task demonstrate how much they care about helping us, as a society, come to grips with the challenges described in this book.

In these times of growing challenges, the difference between working with or against one another can be the difference between food abundance or shortage, environmental preservation or destruction, and even life or death. Politicians must be empowered to make wise decisions. We, the consumers of food, need knowledge so we can wield our influence and purchase power wisely. Farmers should thrive with the best information and practices possible. Funds must flow to wisely chosen research projects that fit into the larger needs of society.

Many scenarios are possible for the future of our food, our world, and its people. Some are disturbing, but others are quite encouraging. Where we end up hinges mightily upon how we address the challenges outlined in the following pages. We can spiral into conflict, failing to nourish the people of our world, spawning illness, suffering, death, and injustice. Or we can thrive together as a global community, ensuring that all human beings have access to sound nutrition for themselves and their children. We sincerely hope that this book leads us, as a society, closer to a conversation focused on facts and shared values that will carry us nearer to this goal.

Chapter 1

Inhabitants of Earth

Brigitte S. Waldorf

The world's growing population is more than a matter of numbers.

"How many children do you wish to have?"

This is the first question I ask in my classes on population. Hundreds of students have answered this over my many years of teaching. The vast majority of responses range between zero and two.

But there is almost always one student in class—without exception it has been, thus far, a male student—who wants to have many kids. His answer of "five or six" inevitably triggers a very interesting conversation on the pros and cons of large family sizes. The expense of raising a child is typically their main reason against many kids. In the United States, these costs are estimated[1] to range between $12,350 and $13,900 *per year* for a child living in a middle-income, two-child, married-couple family, and this does not include the social investments made in children for things like public health and education.

After this, our class conversation generally turns to awareness of and access to birth control, contraceptive failure, and unintended pregnancies. At last, we branch out to the question of how our own individual fertility decisions will impact the world's population growth. It is this, our growing, global population, which is the great matter of concern. Yet few of us think about how our tiny, personal fertility decisions (which don't feel tiny to us at all, coincidentally) contribute to the giant, increasing ticker that measures the quantity of people living on our planet. What even fewer of us think about is that the matter of population is not nearly so simple as how many kids we have or plan on having. There are important follow-up questions we must answer, such as, Where do you plan on living? How old are you going to get? How long can your children expect to live?

It comes as a surprise to many that our personal life choices—beyond how many offspring we wish to have—influence population. Not smoking, eating a healthy diet, exercising regularly, and following all the doctors' advice influence our lifespan. Those extra years we add to our time on Earth will, ultimately, also impact how large the world population is and how fast it will grow.

As I explain to my classes, these choices and decisions, such as eating well or going to the doctor, arise from a place of privilege wrought from a certain socioeconomic status. To decide how many children to have, and how to live, demands a degree of education and economic power. Yet not everyone shares similar privileges. Not everyone can choose. It sounds paradoxical but, as we'll see, *not* having a choice can have an even greater, and sometimes undesirable, impact on the world's population growth.

One area in which this manifests quite keenly is women's reproduction. The United Nations reports that in certain African countries the availability of contraceptives (such as condoms and other means of birth control) is very low. There is a huge, unmet need for family planning.[2] Without contraceptives, having kids "just happens." It is not a surprise that these countries also have the highest fertility rates (or number of births in a woman's lifetime) in the world.

Another space in which we see this is death. With respect to mortality, the desire to live longer is certainly universal. But in many parts of the world, people don't have the luxury to contemplate their health, lifestyles, environmental quality, and long-term well-being. The urgent need to simply survive in the here and now overshadows these concerns. Filling an empty stomach in that exact moment is what counts, especially when you can't be sure of the next opportunity to do so.

Whether wrought by choice or circumstance, these individual behaviors are embedded in a broader context. This broader context is shaped by many factors, such as the environment, economy, and policies. The chapters in this book delve into these issues, and, as you sift through the pages, you can well imagine how some of these challenges influence population levels. Environmental factors such as desertification and soil degradation reduce the amount of food farmers can grow on their land, which in turn diminishes the population that can be sustained. Even more extreme is the complete loss of arable land along the coasts and rivers as a result of climate change–induced sea level rise.

With the broader context taken into consideration, population experts agree we will likely hit about 10 billion people by the year 2050. Yet what does that number even mean? Can we achieve global food security and improve nutrition with so many people on Earth? How can we do so? When will we be able to do so, and what limits humankind's ability to achieve global food security?

Concern about our ability to provide sufficient food for the planet's growing population is at the center of the population debate. The food–population nexus is full of intricate nuances. Issues vary across countries and people. And the complexities only increase as population numbers are subjected to, and contribute to, an uncertain future in a world of elaborate political threads, environmental problems, and climate change. Planning for a future in this complex, multifaceted web of population, food, environment, and climate is a monumental task. The manner in which we confront the uncertainties surrounding future population change underpins the entire food debate.

On the Rise

The world population has been, for the most part, growing ever since the first agricultural revolution about 10,000 years ago. This was when humans began to transition from hunting and gathering to settled agriculture. Growth was slow and steady for the next eight millennia, but for a few calamitous interruptions along the way. The most well known of these is the Great Plague in Eurasia in the fourteenth century. As disease swept the continent, a shocking portion of the population died, with estimates running from 20 to 50 percent and higher in some areas.

But aside from these few exceptions, caused by epidemics such as the Great Plague or prolonged wars, the population growth was very slow and barely noticeable. There were but a few million people at the beginning of the first agricultural revolution. From that point, it took approximately 8,000 years for numbers to grow and hit one billion in 1800. Yet once that first billion was hit, growth skyrocketed. Adding the second billion took 130 years: by 1930, we had doubled our population in less than 2 percent of the time it had taken us to get from the first agricultural revolution to one billion.

Then, in less than a hundred years, we jumped from two to nearly eight billion people (fig. 1.1). As the graph shows, the time needed to reach each consecutive billion became shorter and shorter, plateauing somewhat over the past 50 years. Without giving too much away, since I was born, our population has grown by literally billions. The same can be said for you! Today, 7.6 billion people share our planet. And just this year alone, we will welcome 83 million. That is far more than twice the entire Canadian population alone. The additional 83 million people per year need shelter, health care, education, jobs, transportation, clean air, and clean water. And, of course, food.

It is expected that the threshold of 8 billion will be reached just a few short years from now, at the end of 2022. While I mentioned earlier that experts predict 10 billion by 2050, that is still a rough estimate. Why? It is not simple to predict population growth. Any

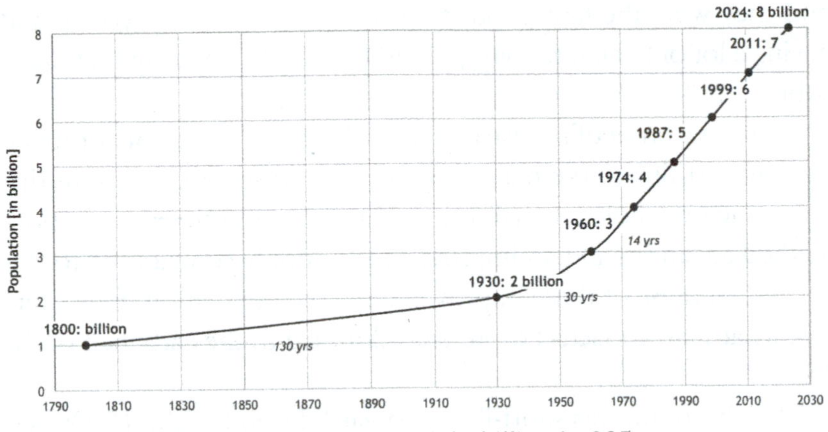

Figure 1.1. From one to eight billion in 225 years.

statement about how the population size will change in the future is filled with uncertainty. These estimates require us to make assumptions about future demographic behaviors and patterns. That is, we need to sketch out scenarios to describe how we will behave.

The United Nations has made projections under different scenarios.[3] First, imagine a scenario where both fertility and life expectancy remain as they were during the recent past. This is the United Nations' "no-change scenario." It is a benchmark for comparison. The result is an astounding 19.7 billion inhabitants on Earth by 2100. This is an additional 12 billion people, or more than two and a half times as many people as currently inhabit Earth. Once we delve into the specifics of providing housing and food for an additional 12 billion people in just 80 years, it becomes quite clear that this scenario would make the twenty-first century extraordinary, to say the least.

Yet we likely won't face such a scenario. Assuming "no change" in demographic patterns would be entirely counter to what has played out in the past centuries. The past 300 years have shown us how drastically demographic patterns can change, and how differently they play out in different places.

To better understand this, let's look at England. England was the first country to go through the demographic transition from high mortality and high fertility to low mortality and low fertility. Simply put, they were the first population to shift from dying young and having a lot of babies, to living a long time and *not* having that many babies.

After the onset of industrialization, England's mortality dropped in large part due to the major decline in deaths from infectious diseases, such as cholera, smallpox, and typhus. Improved sanitation, medical advances, and an improved food supply contributed to this decline. Because of these improvements, the population grew quite rapidly at first. In fact, it more than doubled in the early half of the 1800s.

Yet this trend was short-lived. Toward the end of the 1800s, fertility rates began to decline sharply. In 1871, English women had, on average, 5.5 babies each. Fifty years later, they had an average of only 2.4 babies.[4] This decline happened even *before* the emergence of modern contraceptives, such as the birth control pill in the 1960s.

To better grasp this, imagine the following scenario. Mary, a British homemaker in 1867, is 33 years old and has seven children all under the age of 16. Her neighbors' families are all, more or less, the same size too. She, her husband Frank, and their children all benefit from the better sanitation, medical treatments, and food supply wrought by industrialization. As the years roll past, something happens that in earlier years was nearly unheard of—only one of Mary's seven children dies during childhood. Mary's children soon grow up and have children of their own. Fast forward to the roaring 1920s. Mary is still alive, now an old woman, and is surprised that her children and grandchildren have kept their families so small. Not one of her progeny has more than four children, and some don't even have any at all!

Mary lived through what we now understand to be a common demographic shift for high-income countries. Today, England has long since completed this transition. Life expectancy is very high, at 81 years, and fertility is quite low, at 1.9 babies per woman. This is even below the replacement level of 2.1 babies, the number of children needed to keep a population stable.

All high-income countries[5] over the past century and a half have followed this very same trend at slightly different times. It is a transition into what we call a "low mortality/low fertility regime." In some countries today, the declines are even more drastic than what we see in England. Take Japan, for instance. The Japanese can expect to live two years longer than the British, and Japanese women have an average of only 1.4 babies.

To many, this initially feels counterintuitive. As the standard of living rises, and there is a lower chance of mortality, we technically have more money to raise kids and we don't have to deal so frequently with the horrors of dying young, or watching children die young. Nonetheless, we have fewer kids. And middle-income countries are well on their way to following this same trend. World Bank data[6] suggest that life expectancy in Mexico climbed from 57 years in 1960 to 77 years in 2014. Due to successful government-sponsored family planning programs, during that same time period Mexican fertility dropped from 6.8 to 2.2 babies born per woman.

Similar developments are being observed in lower-middle income countries. In Myanmar, for example, life expectancy increased from 43 to 66 years between 1960 and 2014, whereas the number of babies born decreased from 6.1 to 2.2 per woman. In low-income countries, life expectancy has already increased whereas fertility is only starting to decline. In Mozambique, for instance, people now live 55 years on average, up from 35 years in 1960. Yet the number of babies born per woman only slightly dropped so far, from 6.6 babies to 5.4.

These changes are the reason why the "no-change scenario" and its prospect of 19.7 billion people in 2100 is such an unrealistic outcome. It is far more realistic to think about scenarios that allow for declines in both mortality and fertility, especially the latter. The issue is, what are reasonable assumptions about the speed of the mortality decline? And what is a good estimate for the speed of fertility decline in the middle and low income countries?

Given these complexities, in addition to the no-change scenario, the United Nations also projects other scenarios for the world's future population that take into account varying degrees of declining mortality and fertility. The chart here shows three of these projections (fig. 1.2). Each makes different assumptions about the fertility decline. They are labeled as low, medium, and high (as well as the no change projection). The projections start in 2015 and extend to the end of the century. As you can see, differences among them emerge quite early. After just 20 years, the projections for high fertility and low fertility already differ by almost one billion people.

In the low-fertility scenario, one in which population experts track what they guess will be the lowest probable rates of fertility and mortality, the population never even reaches nine billion. In fact, it already starts declining by 2053. It further projects that by 2100 the population will be even smaller than it is today.

Yet in the high-fertility scenario, where researchers assume the higher end of probable fertility and mortality, the projected population figures exceed those of the no-change scenario for the first 62 years before dipping down below and reaching 16.5 billion in 2100. The medium-fertility scenario, the most commonly cited scenario, projects 9 billion people at the end of 2036, 10 billion at the end of 2054, and 11 billion in 2088. It is not as extreme as the high-fertility and no-change projections. This medium scenario also predicts that population growth will slow substantially as we reach the end of the century. In fact, its growth rate during the last year of the twenty-first century is close to zero (0.09 percent), hinting that population growth will level off in the twenty-second century, or even come to a total standstill.

A note of caution, however. The thing that makes population projections so tricky is all the uncertainty, and the uncertainty grows the further we look down the line. A disease outbreak, for example, can throw these projections completely off kilter. HIV and AIDS have, in some countries, slowed or even stopped the downward trend in mortality. Other recent examples include the Ebola outbreak in western Africa, and the emergence of a deadly flu virus similar to the Spanish flu H1N1 influenza virus that spread in 1918–19 and killed millions of people. Public health officials are particularly concerned about this. Such a virus can, in our globalized world, easily

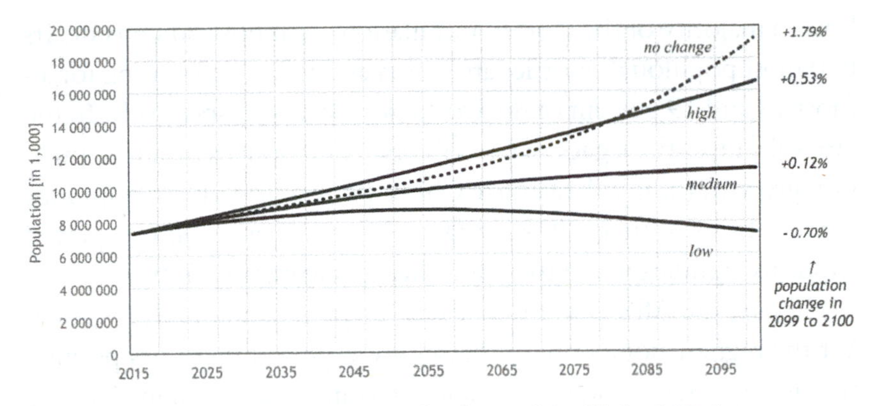

Figure 1.2. Population projections of the United Nations.

cause a pandemic with devastating consequences, including large population losses. Don't panic yet though, because successful vaccination programs can and have lowered mortality worldwide. Improvements in quantity and quality of the food supply do the same.

Yet it is not just mortality, or unexpected deaths, that affect projections. There are unforeseen fluctuations in fertility. China recently loosened its one-child policy, allowing women to have a second child. There are more than a quarter billion Chinese women of reproductive age (15 to 44 years old). Imagine if all of them decide to have an additional child. This would be an immediate and immense boost to the population, and it doesn't stop there. These additional offspring may also decide to have two children, thereby further multiplying the initial boost 25 years or so down the line.

Given that it is impossible to include all such unpredictability into projections, the United Nations updates its estimates every two years. It's good to check in. And as frustrating as it may be, there is no clear, obvious answer as to just how many people will grace our Earth in 2050, let alone 2100.

Location, Location, Location

Assuming that there *will* be 11 billion people on Earth in 2100, as the medium-fertility variant suggests, where will they live? Asia currently hosts a majority of the world's population, at almost 60 percent. Its two most populous countries are China and India, accounting for almost 18 and 19 percent of the world population, respectively. In the years ahead, as the legacy of China's one-child policy curtails population growth, India will overtake China as the world's most populous country by as soon as 2024. In fact, China's population is expected to shrink significantly in the years ahead, starting in 2030.

Today, the African continent has fewer inhabitants than China. Yet the biggest population increases are expected there. Its population is very young, and the prevailing fertility rates are high. This is a perfect recipe for very fast population growth. In fact, its population is projected to rise from 1.2 billion to almost 4.5 billion in 2100.

In Niger, for instance, women today have on average more than six babies. Add to this the fact that young girls under the age of 15 make up a quarter of all inhabitants. This large cohort of young girls will be adult women and mothers very soon, and, even if these girls have only three or four children on average,[7] they will contribute substantially to Niger's population growth. In fact, Niger's population is expected to double in size in the next 19 years, from 21.5 million in 2017 to 43 million in 2036.

This pattern of young populations combined with high fertility levels is typical for African countries. Figure 1.3 demonstrates where current populations are distributed and projects estimates of where the world population will live in the future. As you can see, the continent of Africa is set to rise considerably, overtaking all other regions by a long shot.

The other extreme is Europe. Compared to Africa and Asia, Europe has very few people and makes up only 10 percent of the world population. In the years ahead, its population will shrink further. By the end of the century, Europeans are projected to account for less than 6 percent of the world population. Even today, Italy's slow population growth has led to the government's proclamation of a "Fertility Day" (we can only imagine what that encourages). In Germany, fertility levels have been below replacement since the 1970s, despite many government policies geared toward lowering the costs

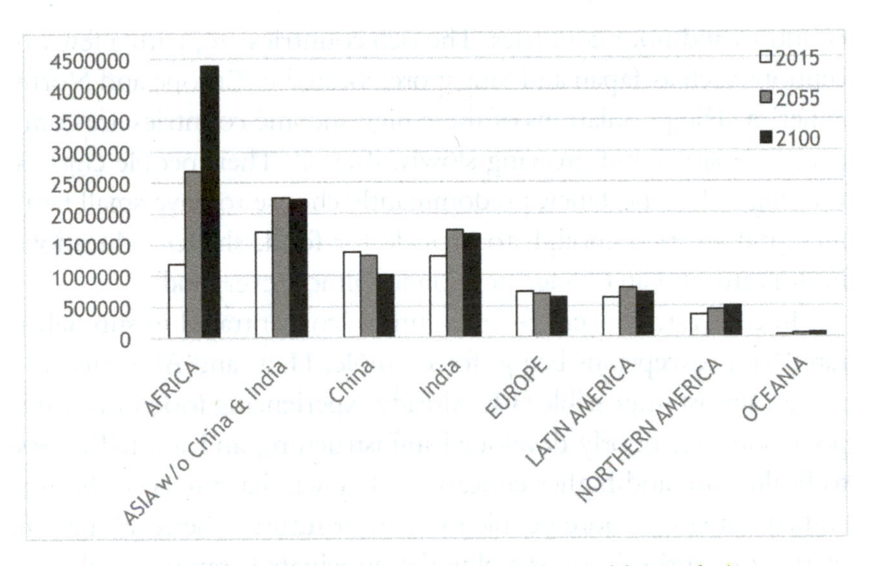

Figure 1.3. Projected distribution of the world population.

of child rearing, including subsidies and generous maternity leave arrangements. And the population itself isn't just shrinking, it is also aging. Not surprisingly, half of Germany's population is older than 47 years. In comparison, half of Niger's population is younger than 15 years.

Northern America and Oceania also start with small populations, but their growth rates are projected to be somewhat higher in the years ahead. Yet, unlike in African countries, the main countries of these two regions (the United States, Canada, Australia, and New Zealand) grow because of immigration, not because of fertility. Immigration is also what keeps these countries younger than their European counterparts. The median age in Australia and the United States is, for example, almost 10 years younger than in Germany.

Divides and Dividends

To understand the spread of inequality, look at population. It reflects inequality quite starkly. The distribution of people across continents signals a particular kind of demographic divide, one between rich countries and poor countries. The rich countries are, with a few exceptions such as Japan and Singapore, located in Europe and North America. The populations of these high-income countries are comparatively small and growing slowly, if at all. Their people enjoy a very high life expectancy, predominantly choose to have small families, and are rich enough to provide for food, shelter, education, health care, and an old-age pension for almost everybody.

In contrast, the very poor countries, concentrated in sub-Saharan Africa (exceptions being, for example, Haiti and Afghanistan), face an almost impossible task. Already experiencing food insecurity, poor housing, poorly developed infrastructure, and limited access to health care and higher education, they will have to feed, house, and educate even more people in the near future. The combination of these countries' poverty plus the anticipated, rapid population growth is what makes this task so difficult. Imagine that the United States would grow so fast that, in 25 years, it would have to provide housing, food, health care, education, and infrastructure for twice as many people. That is, for 640 million people rather than just 320 million. Even a rich country like the United States would not be able to do that, let alone a poor country like Niger.

In cases like that of Niger, where population growth is expected to be so rapid that the population will double in 25 years, growth can be slowed if fertility is drastically lowered. Many countries in Asia and Latin America have done so. And, if the fertility decline happens fast enough, these countries may actually enjoy what is called a "demographic dividend." South Korea, for example, reduced its fertility rapidly. In 1960, South Korean women had 6.2 children on average. Some 25 years later, they only had 1.5 babies. As a result of this rapid decline, the population's age structure changed such that the share of economically active population (age 15 to 65) became very large, and the share of dependents (the young and the old) became quite small. Such an age composition offers immense economic advantages. Society has few expenditures for educating the young and caring for the older population. However, the demographic dividend is often of short duration and is typically followed by an aging problem, especially if fertility remains low or drops even further. In fact, in South Korea the fertility continued to drop to 1.2 children per woman in 2014. The South Korean government is now, not surprisingly, encouraging larger family sizes.

The question that remains is whether the poor and fast-growing countries of sub-Saharan Africa can follow these examples and similarly reduce fertility. Knowledge of, access to, and use of contraceptives are seen as necessary for the reduction. Yet given the widespread, unmet need for family planning in this region, a rapid fertility decline seems unlikely. A complementary trajectory, however, involves education—especially girls' education. Rising educational attainment levels are associated with lower fertility. Yet investing in education is likely to be a very slow path out for these high-fertility regimes. After all, education is expensive and happens over a time span of years. Poor countries may not have the resources to make such investments.

Migration

Many of us can look back over our family's history and see movement. Our grandmother might have come from Poland, or our grandfather from China. Perhaps our great-great-great-great aunt was the descendant of a slave who was forcibly relocated from West Africa to Alabama. Or perhaps we ourselves are immigrants. I was born in Germany and now live thousands of miles away from my birthplace, in my new home country of the United States.

As we have seen in countries like Italy or my birth country of Germany, if we lower fertility rates enough, population growth halts altogether. At a regional and local level, however, population growth can *also* be slowed by emigration. People can leave their country and move elsewhere. Germany's population, for instance, dropped by one when I moved to the United States.

Mass emigrations that have a big impact on population distribution (unlike the more isolated cases such as my own) have happened throughout history and in the more recent past. When England's population was growing rapidly during the eighteenth and nineteenth centuries, a good deal of the population left England and settled in North America, Australia, and other countries throughout the Commonwealth. Leaving home and finding a better life in the New World of the Americas and Oceania was also an option for families in many other countries across Europe during their times of rapid population growth.

These receiving countries were initially formed through immigration and have, ever since, developed policies to regulate additional immigration. Today, these policies clearly define who is allowed to enter the country, under what conditions, and for how long. Most immigrants enter the receiving countries under the umbrella of family unification or by taking advantage of skill-based entry permits.

Today, however, relieving the population pressures of African countries through emigration faces major constraints. Entering the United States via family unification, for example, is very limited because few Africans have immediate family in the United States. The ancestors of existing African Americans in the United States were almost all forcibly taken elsewhere and enslaved. Slavery meant that their knowledge of family history in Africa was castrated. They never had the chance at unification, even after slavery officially ended.

Moreover, the vast majority of Africans are poorly educated. This means that they can't offer the kinds of skills that would qualify them for skill-based visas. And when highly educated Africans do leave, the associated brain drain frequently further impedes development. The Migration Policy Institute[8] recently reported that, in the past, many doctors and nurses who were born and trained in Malawi left the country, thereby contributing to severe shortages of health care services. This example hints at a pattern seen quite often in the migration from low-income to high-income countries—those who leave are often those who are most needed for the country's development.

What's more, European countries and Japan categorically close their doors for what they call "economic migrants." Some European countries have welcomed large numbers of asylum seekers. However, none of them provide a legal path of entry for those simply seeking a better life. The reality, nevertheless, is that increasing numbers of young Africans are taking a chance and trying to enter Europe, often under quite dangerous circumstances, for the hope of something better.

Urbanization

There is a much higher probability that your grandparents lived in the countryside than you do today. That is because movement doesn't just take place across country lines. We also move around within our countries, from low-populated areas to high-populated areas, or vice versa. Today, we see a lot more people moving *toward* high-populated areas rather than the other way around.

A much greater share of the population in fast-growing countries now takes part in what we call the "rural-to-urban migration streams." Moving to cities is certainly not new, though. England's industrialization in the eighteenth and nineteenth centuries was tightly linked with the migration of surplus labor from rural areas to the new cities forming around England's coal mines and textile industries. In fact, all developed nations became increasingly urbanized as they transitioned from agricultural to industrialized economies. The Americas and Europe are the most urbanized regions of the world, with Asia catching up quickly. In fact, we have seen unprecedented migration flows from China's rural areas into Beijing and Shanghai, as well as into the more than 70 other fast-growing Chinese cities with over a million inhabitants.

By now, the majority of the population, about 55 percent, lives in urban areas. This is up from about 2 percent at the beginning of industrialization. In that sense, the story of population growth is also a story of urbanization. And, as the famous urban economist Edward Glaeser points out, urbanization and the growth of cities is a story of the "global transition from poverty to prosperity."[9] Indeed, cities provide a wealth of opportunities. There is a diverse range of jobs in many sectors of the economy. The urban wage premium adds a nice bonus. The extra prosperity creates a difference in consumption patterns. For example, compared to their rural counterparts, urban residents consume much more meat. The lifestyle and (relative) affluence of urban residents also imply that they use more electricity and buy more durable goods. Cities also offer services that are often lacking in rural areas or are only available in a rudimentary form, such as education, health care, entertainment, a variety of consumer goods, and access to better infrastructure, from electricity to roads. The fast growth of cities makes sense in this context, and we expect most of the world's future population growth to be in urban areas.

About half of the urban population lives in small and medium-sized towns and cities with fewer than 500,000 inhabitants.[10] Forty percent live in the 512 cities of the world that have more than a million inhabitants. Most fascinating of that group are those living in the 31 megacities on Earth. Megacities have more than 10 million inhabitants. Among them, Tokyo is the largest, with a population of 38 million, followed by Delhi and Shanghai. In total, China has six megacities, India has five, and most of the remaining megacities are located throughout Asia, Africa, and Latin America. Europe has only four—Istanbul, Moscow, Paris, and London—and the United States has just two megacities, New York and Los Angeles.

We expect more megacities to emerge within the coming years, Bangkok in Thailand and Luanda in Angola, for instance. However, where rapid urbanization does not keep pace with economic development, pockets of extreme poverty emerge. Squatter settlements and slums are part of many large cities. The favelas of Rio de Janeiro, for example, came into focus during the recent Olympics.

While cities offer all the advantages of urban agglomerations, as listed earlier, these same large agglomerations also create problems like congestion, unbridled spread of infectious diseases, pollution, and the urban heat island. Temperatures in cities tend to be higher than in the surrounding rural areas, with a difference of about 2°C. This urban heat island (akin to a localized version of climate change) traps pollutants that, in extreme cases, trigger smog alarms. Such smog alarms are well known across cities like Los Angeles, Beijing,

Paris, and Mexico City. Urban pollution has severe consequences for people's health, including higher death rates. Yet protective measures (e.g., driving restrictions) are a disruption to public life and can trigger economic losses.

On a broader scale, many cities face the consequences of climate change. Being located along the coast creates wealth from trade and was traditionally an advantage for cities like Amsterdam, New York, and Shanghai. But the coastal cities are now exposed to rising sea levels as glaciers melt at a rapid pace. Flooding in Miami is already starting to affect homeowners. Large-scale displacements and environmental refugees may very well be the headlines of the not-so-distant future.

Our world's cities are also not immune to natural hazards. Moreover, in cases of disaster, cities pose challenges that, particularly in the case of megacities, can take on dramatic proportions simply because of their sheer size. There can be floods, tornadoes, volcanoes, earthquakes, and hurricanes. The 2004 tsunami in Banda Aceh, the 2005 hurricane Katrina over New Orleans, the 2010 earthquake near Santiago de Chile, the 2011 earthquake and subsequent tsunami around the Fukushima power plant, the 2011 earthquakes in Christchurch, New Zealand, the 2012 tropical storm Sandy in the greater New York area, and the 2017 hurricane Harvey in Houston are recent examples of natural disasters' devastating impacts on highly urbanized areas. The devastations are especially severe in cities where urban dwellers have insufficient protection.

The Puzzle Picture

In putting the pieces together, we clearly see that the world's population is growing and will continue to do so in the years ahead. This population growth is, and will be, unequal across countries and continents. Africa and Asia will take the lead. The populations of rich countries, including Europe, Canada, the United States, Japan, Australia, and New Zealand, will lag behind or even shrink. The driving force behind this unequal population growth is the high fertility in

Asian and African countries, and the lack thereof in the rest of the world.

This demographic divide has been, and will be, shaping the world population. The small and slow-growing populations of the high-income countries will struggle with the consequences of aging and, in some cases, even population decline. The fast-growing populations of the poor countries will struggle with the need to create jobs, housing, infrastructure, education, and health care for the younger population.

Chances for a redistribution of the population across the demographic divide are limited given factors we explored earlier. What we will likely see, however, is a rapid redistribution of the population from rural to urban areas. In the high-income countries, urbanization is already very advanced, and in many places, more than 80 percent of the population already live in towns and cities. In middle-income and low-income countries, urban areas are growing faster than the rural ones. The impacts of this urban growth are substantial. Consumption patterns shift, pollution increases, water resources are used up, and the environment is indefinitely changed, both locally and globally, through the consequences of climate change, from rising sea levels to desertification. These effects in turn have health implications ranging from higher incidences of respiratory diseases to higher infant mortality rates.

A question that has long dominated the population debate is whether Earth can actually sustain such a large population, be it 11 billion or even higher. At the end of the eighteenth century, when England's population started to grow and urbanize rapidly, the English clergyman Thomas Robert Malthus argued that unchecked population growth would lead to food shortages and famines. During the second half of the twentieth century, when the world population grew rapidly, the so-called neo-Malthusians (e.g., Anne and Paul Ehrlich) rephrased his argument in modern terms, warning not only of global food shortages but also of using up the nonrenewable resources like oil and gas.

There were, however, other voices that opposed the doom scenario of an exploding population bomb. Ester Boserup, a Danish economist, argued that population pressure leads to intensification, new technologies, and eventual development and prosperity. I find that I, myself, am also somewhat of an optimist. Our world imploded neither back in the late 1700s as Malthus predicted, nor in the latter half of the last century as Anne and Paul Ehrlich so direly foretold. Yet we also cannot sit around and do nothing, just hoping for the best.

The many challenges addressed in the following pages of this book, such as land use, environmental issues, and trade, are hints of the complexities we will face in the years ahead. Evidence thus far suggests that rising prosperity as well as investments in efficiency enhancing food technology and human capital will be key in reducing vulnerability and ensuring sustainability. The tragedy is that the low-income countries, who have the highest population growth, have themselves the least wealth and access to key knowledge. They have also contributed the least to the negative consequences of environmental change, yet they may be hardest hit. And they do not have much wealth and technology to devote to mitigation.

The question of population sustainability, and Earth's carrying capacity, thus also becomes a question of bridging this demographic divide. After starting the lecture with the question, "how many children do you wish to have?" I perhaps ought to ask my class at the end of the lecture, How will your fertility decision influence the demographic inequalities of the world?

Chapter 2

The Green, Blue, and Gray Water Rainbow

Laura C. Bowling and Keith A. Cherkauer

Our precious, powerful, and oft-overlooked commodity

Laura often starts her lectures on global water resources with a photo of a river in flood. The picture highlights a murky expanse of water that ripples and flows as it washes away the nearby vegetation. The river is vast and full, expanding well beyond its banks. A small community is just visible on the horizon beyond.

She then asks the class if the community shown is experiencing water scarcity. Realizing it is probably a trick question, most don't answer "no." But most also don't understand why until she explains what water scarcity really means—situations where the supply of water, with a good enough quality for the intended use, is low relative to the amount that is actually needed. The community showcased in the photo may very well be in a water-scarce situation. Why? During floods, most surface water is not safe to drink given how many extra contaminants have been swept along with the flood waters. This community has a lot of water right now, but it is of such poor quality that no one can drink it.

These types of simplifications (there is a flood, so there is enough water, right?) are often the case when we think about our global water resources. Water in the Earth system moves in a cycle, the hydrologic cycle, in which water from Earth's surface is recycled by evaporating from the soil, a lake, or an ocean in one location and then transported through the atmosphere before it falls down again as rain or snow somewhere else. This constant recycling means that the volume of water on Earth is approximately constant over time. So technically, on a global scale, there is no such thing as "running out" of water. The challenge lies in the fact that there are many places where we do not have high-quality freshwater when and where we need it. This can happen everywhere, including the United States (fig. 2.1).[1]

We need water to live, we need it to grow our food, and we need it to produce the electricity and fuel that will transport that food to our doors. We use water in the shower. We use it to wash our clothes. It nourished the cow that turned into our burger. It fed the crop of corn that later became our favorite chips. Despite its critical importance to all life, including our own, many of us often take water for granted, assuming we will always have access, without consideration as to how our use might affect the quality and quantity of water available for others. Yet, as we will see throughout this chapter, our water supply is far more nuanced, precarious, and necessary than most of us have considered.

We mentioned water supply in that last sentence. That is because all of the water we use comes from somewhere. It is supplied to us. The source of that supply is what we call a water resource. Water resources are all the sources of water that are useful or potentially useful to humans. Although this is something of a dry definition (pun unintended), used for qualitative purposes, it does in fact encompass all of the water that enables us to live the lives that we do, including what we eat. Some water resources we can easily see, such as the

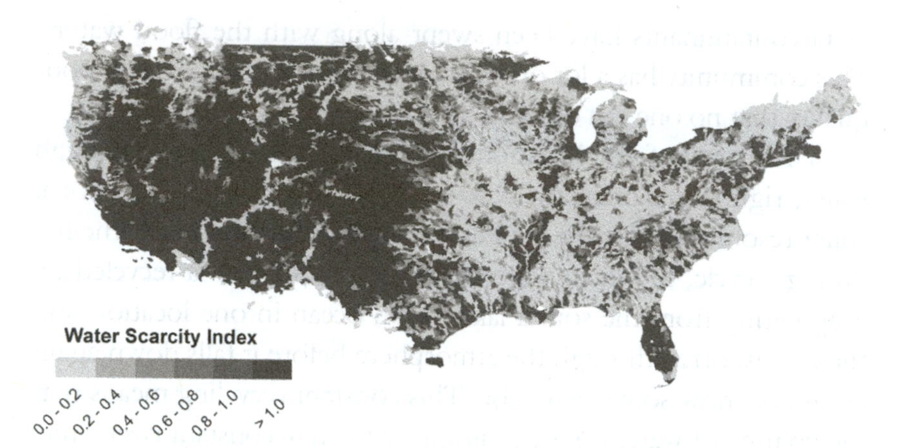

Figure 2.1. Water scarcity in the United States. Areas with a water scarcity index (WSI) greater than 0.6 are at risk.

Ohio River that flows throughout the year (the creek in your backyard that dries up every summer is not considered a water resource).

Before delving too much further into the issues surrounding water, it helps to understand some basics about these water resources themselves. There are two major categories of water resources, surface resources and groundwater resources. Surface resources are those that, as the name makes rather obvious, appear on the surface of the earth. These are perennial rivers (those that flow year-round, unlike that creek in your backyard), lakes, and artificial reservoirs created by damming rivers. Groundwater resources include underground aquifers that are saturated with water that can be pumped to the surface.

Surface water resources, given that they are pretty easy to find and figure out how to use, have historically been developed first. But rivers and streams have the unfortunate habit of varying dramatically in volume in response to weather events. If there are a number of heavy rains in a row, they rise. If there is a period of no rain at all, they may dry up. Surface water resources are not particularly stable sources. Many rivers experience their lowest levels during the growing season, just when farmers most need to irrigate. To combat this, people frequently use what are called surface water reservoirs to help buffer the unpredictability. They store surface water so it can be available during the dry period when it is needed the most. Dams are, of course, such an example. In the United States, the Bureau of Reclamation in the west and the Army Corps of Engineers in the east led the charge during the heyday of dam building in the 1950s. Today, there are over 90,000 dams in the United States.[2] The amount of water that dams on the Colorado River system alone can store is over 77 million acre-feet (an acre-foot is a unit of volume that indicates the depth to which irrigation water can be applied to an acre of land, so with all of this water, 77 million acres could be irrigated to a depth of 1 foot when all of the reservoirs are full). The current trend in reservoir building in China and elsewhere, such as the controversial Three Gorges Dam on the Yangtze River, is mostly focused on hydropower generation. However, climate change has made surface storage increasingly attractive for places that have not historically relied on it.

The handy thing about groundwater resources is that they are often available in locations where surface water is not. The use of groundwater took longer to grow given that it is harder to get to, but the development of inexpensive tube wells (a pipe and pump) in the 1950s changed things. It led to a rapid tapping of groundwater resources, such as in Southeast Asia, where it was used for irrigation. However, the original push was not only for irrigation but also to promote public health. Residents were encouraged to drink from the cleaner groundwater supply rather than using the surface water, which was polluted from erosion and fertilizer runoff. Unfortunately, water can be a tricky thing, and this policy backfired in some regions. For example, in Bangladesh, after many years of encouraging the development of groundwater resources, the government slowly realized that the high naturally occurring arsenic content of that particular groundwater source was actually poisoning the population.

Although civil engineers are commonly thought of as those who build bridges and roads, there are also dedicated water resources engineers (a type of civil engineer) who focus on designing systems and equipment specifically for water. These engineers have historically relied on the two types of water resources we explained thus far, surface and ground, when developing infrastructure. These are commonly known as blue-water resources. Increasingly, however, the international development community is discussing "green water" as a third water resource. "Green-water resource" refers to the moisture that seeps into the soil during rain and is sucked up by plants for growth. Additionally, there is water that has been used by humans that does not contain sewage, such as treated wastewater or wastewater from sinks and showers that could be used for irrigation. This is considered a gray-water resource.

With this in mind, it is time to turn toward the critical role of water in the production of our food.

More Crop per Drop

It comes as no surprise that plants need water to grow. They simply cannot live without it. The same can be said for all the different types of animals that we eat—cows, chickens, and more (not to mention fish!). In addition to the water that those animals drink, they also need to eat food, derived from plants, to stay alive.

Remember learning about photosynthesis back in elementary school? Water is used by plants as part of photosynthesis to build biomass, the carbohydrates that form the basis of the plant roots, leaves, and stems. But by far, the greatest amount of water used by a plant does not become a part of the plant at all. Instead, it passes through the plant and is released back into the atmosphere as water vapor in the process called transpiration.

Imagine a two-passenger roller coaster that loads the passengers from both the left and right sides of the ride. The water line is one side, and it starts in the plant roots where water is drawn up into the stem and eventually to the plant leaves by suction pressure. The other line, representing carbon dioxide, has a much shorter queue. That line extends only as far as the stomatal openings on the plant leaf, where carbon dioxide enters whenever these pores are open.

When the "passengers" get to the front of the line, six water molecules and six carbon dioxide molecules enter the car, side by side, and go together on the roller coaster ride of photosynthesis out into the sunlight. It is such a wild ride that, when they get back to the station, they are all entangled. Six oxygen molecules now exit through the gate that the carbon dioxide used to get on the ride, leaving behind a new carbohydrate molecule that will be used to build onto the roller coaster (biomass).

Without water, the plant cannot build the necessary biomass. But there are a lot of water molecules in line that never get on the roller coaster. They get to the front of the line and exit immediately through the same gate (stomatal opening) that lets carbon dioxide in and oxygen out. Far more water molecules skip the ride than actually ride the coaster to create biomass. The water use efficiency, often defined as the mass of carbon assimilated per mass of water lost in transpiration, quantifies this exchange. Typical water use efficiencies vary around 0.00005 moles carbon assimilated per mole of water transpired. In other words, up to 20,000 more water molecules skip the ride and move directly to the atmosphere than are actually used by the plant to create biomass.[3]

This is important. It affects how much food we can grow for any given amount of water. The amount of water needed for carbon assimilation (to create plant biomass) is fixed, and we can't change this. But where we *can* work to increase productivity is by adjusting the amount of assimilated carbon that goes toward the parts that we eat, rather than the oftentimes inedible roots, stems, and leaves. Take, for instance, a stalk of wheat, most of which we don't consume, such as the long stem, roots, and leaves. All we really want are the tiny kernels waving in the breeze right at the very top. That is only a very small portion of the plant's biomass. This is called the crop's harvest index. It is the ratio of grain to total biomass. Adjusting it is an active area of research in plant breeding and genetics.

There is an additional catch-22 here. Transpiration water loss is a function of the plant physiology, but it is also very sensitive to climate conditions. A plant will lose more water to transpiration in a dry climate, whereas it will lose less in a high carbon dioxide environment. In other words, it takes more water to grow crops in the very areas that are already water scarce. Additionally, in every field there is always some water lost simply due to direct evaporation from the soil. This so-called nonproductive evaporation is water that never even passed through the plant to begin with. It is wasted from a food production point of view from the get-go.

We can see that, during crop growth, a huge quantity of water is consumed that doesn't actually stay in the plant. This led to the development of a very interesting term, "virtual water," which was coined by Tony Allan as a way to discuss water scarcity.[4] To summarize, every food product we eat contains some volume of "invisible water" that was used during its growth but that is no longer contained in the actual food product itself. The beef patty on your hamburger, for instance, doesn't contain even close to the amount of water that was necessary for its existence on your plate (quite a bit of feed was necessary for that cow to grow). The same can be said about your pile of rice. Much of the water is returned to the atmosphere as water vapor while the crop is being grown.

Allan suggested the use of a virtual water budget as a way to promote food security for water-scarce countries, such as Saudi Arabia. This water budget showed that it is far more practical and economical *not* to transport water to the interior of the country to grow food locally, but rather to transport food that was grown in a more water-abundant environment. Two researchers, Chapagain and Hoekstra, expanded the concept of virtual water.[5] Today, it is the basis for the "water footprints" of hundreds of food and consumer products. The water footprint is the volume of water needed to produce the final product. In the case of food, without question the majority of the water footprint is the water used by the crop for transpiration and nonproductive evaporation. It is not found in the final food products; no one sees the amount of water that went into creating the food on their plate. Thus the virtual water budget and water footprints were created to provide a broader picture.

There are online water footprint calculators. They give broad estimates of the amount of water needed to produce different types of foods. Estimates can vary from the 462 gallons of water required to produce a quarter-pound hamburger (just the meat patty), to the 308 gallons for an equivalent calorie content in a bowl of lentils.[6] When we talk about water footprints with undergraduate students, their first instinct is that the water footprint for the hamburger must be so much higher because of the water needed to care for the animals. But just as direct consumption of water by plants is very small, the water used for direct animal needs is just a "drop in the bucket." The majority of the water used (98 percent) is the water needed to grow the food that feeds the cow.[7] For an adult cow that is on the order of 30 pounds of hay per day.[8] So only about 2 percent of the 462 gallons of water required to produce the hamburger is directly consumed by or used on the animal.

Water footprints are typically broken into three categories based on the source and use of water. The first is the green-water footprint, which is water that falls on a location as precipitation and is stored in the root zone of the soil and is directly available to plants. Next is the blue-water footprint, which is water withdrawn from surface or groundwater supplies. This can take many forms, but most relevant to food production is its use for irrigation. Finally, gray water is the water used to dilute pollutants to meet water quality standards.

Returning to our story on food production, the water footprint for beef is always quite high, but when considering just the blue-water footprint, that of a grazing animal relying on rain-watered pasture will be much smaller than that of a confined animal eating irrigated corn. A higher green-water footprint is much more sustainable than a high blue-water footprint. Thus not all beef is created equal in terms of water use. As such, we typically point out that these global water footprint calculators are quite general, relying on national average production practices. These often mask the potential for consumers to make certain choices to reduce their own water footprint by substituting similar foods that were produced more efficiently.

For example, picture the array of tomatoes in a grocery store.

Heirloom, on the vine, greenhouse, organic—which is the best tomato to choose? Since your two authors live in a water-abundant area (Indiana), in the summer a locally grown field tomato is always going to win in terms of both taste and water footprint. But in the off-season, we will often go for the greenhouse tomato because the efficiency of water recycling in these systems contributes to a lower water footprint than the tomato that is vine ripened outside in a water-scarce environment.

When Keith teaches this part of his class to agricultural engineers, he conducts a little exercise to evaluate the water used to grow an apple. From the US Department of Agriculture's National Agricultural Statistics Service, the average amount of apples grown in the United States between 2011 and 2014 was about 31,000 pounds per acre. Growing 1 pound of apples requires about 80 gallons of water, which works out to 57 gallons (7.6 cubic feet) of water for each square foot of apple production (or 91 inches of water per unit area). Average annual precipitation is 37 inches for the United States, with 32 inches in Michigan and 8.4 inches in eastern Washington (two primary apple-growing regions). Neither of these regions can support apple production based solely on the precipitation—that is, the green-water resource that falls on the trees and soil. Instead, to maximize production, they must rely on irrigation from blue-water resources and expand their water footprint.

Green-Water Resources Key to Sustainable, Rainfed Agriculture

So plants need water to grow. But how do crops get water when and where they need it? Indeed, the water does sometimes magically fall from the sky, but rain doesn't always fall according to our schedule. This is why humans developed irrigation, which we'll discuss later in the chapter.

The rain that falls from the sky and lands directly on the soil, settling there, and feeding straightaway into the roots of plants, is the most direct and sustainable way of delivering water. This type of agriculture is referred to as rainfed, and it uses a green-water resource. As already mentioned, among the various types of water resources, this is the one that is the most efficient and ecological.

However, as alluded to, the single biggest challenge in managing rainfed agriculture is the variability of rainfall. This variability controls whether or not a crop will fail or thrive. This is due to its impact on whether or not a location has a pronounced wet season and dry season, or receives fairly equal amounts of precipitation in each month of the year, or whether precipitation comes in dramatic convective thunderstorms or arrives in slow, soaking warm fronts. These are all an inherent part of the regional climate, just as the average annual amount of precipitation and the average annual temperature can be used to define a location's climate.

Many people assume that the midwestern Corn Belt has a pronounced rainy season because our soils dry out so much later in the growing season. But this is actually a reflection of the extreme amount of moisture lost to evapotranspiration during the long, hot summer days. The monthly distribution of precipitation is actually fairly equal. Managing the annual dry season is a routine part of rainfed agriculture. But the true challenge comes from interruptions and changes to any given area's seasonal pattern, something that is increasingly playing out around the world due to climate change, as chapter 4 delves into in greater depth.

If the annual cycle of wet and dry periods represents climate, drought is the departure from these normal conditions to a drier state. Neil deGrasse Tyson, an American astrophysicist and science communicator, has a YouTube video in which he is walking down a beach with a dog on a leash. While he walks a fairly steady path down the beach, the dog runs back and forth in front of him, to the end of his leash and back. Tyson explains that his footprints represent the climate (the long-term average values), while the footprints made by the dog, veering back and forth relative to his path, represent weather.

Some locations have greater swings in natural weather variability than others. Imagine a husky with a longer leash and more energy compared to a basset hound on a short leash. This can help explain why we hear more about drought in some locations rather than others. Meteorological drought occurs when precipitation is lower than normal conditions for the time of year. While any location can experience drought, a location represented by the husky on a longer leash (experiencing more dramatic and rapid variation), may be considered more drought prone because dry conditions can be more severe and occur more frequently.

For example, much of Southern and Central California receives less than 20 inches of precipitation a year, based on long-term (1971–2000) average precipitation measurements. This is already a pretty low amount, which is why the area experiences long-term water scarcity. Yet, in the Los Angeles area, 8 of the last 10 years have had below-average precipitation for what is normal in the city. This location has historically had more years per decade with below-average precipitation than with above-average precipitation. What does this mean? The years that are wet are really wet. But most years are dryer than average. So while the husky's footprints are still part of the normal variability of a region, there is a big swing between wet and dry. Areas like this, with the greatest changes year to year, present the biggest challenge for rainfed agriculture.

As you can see from figure 2.2, the number of wet years in Los Angeles varies radically from decade to decade. It can just as easily be two wet years per decade as it is six years per decade. It is quite a challenge to plan out an effective agricultural strategy with so much uncertainty.

Drought is by definition short term, whereas water scarcity is long term. Drought is a departure from normal. Eventually, the dog reaches the end of its leash and gets pulled back to its owner (or average conditions). A drought cannot last for decades, since typically we define "normal" conditions in terms of a 30-year period; thus a 30-year drought would suggest a "new normal."

So how can we manage this uncertainty? We can manage our green-water resource better. One such way to facilitate this is, interestingly enough, improving soil health. Water that runs off over

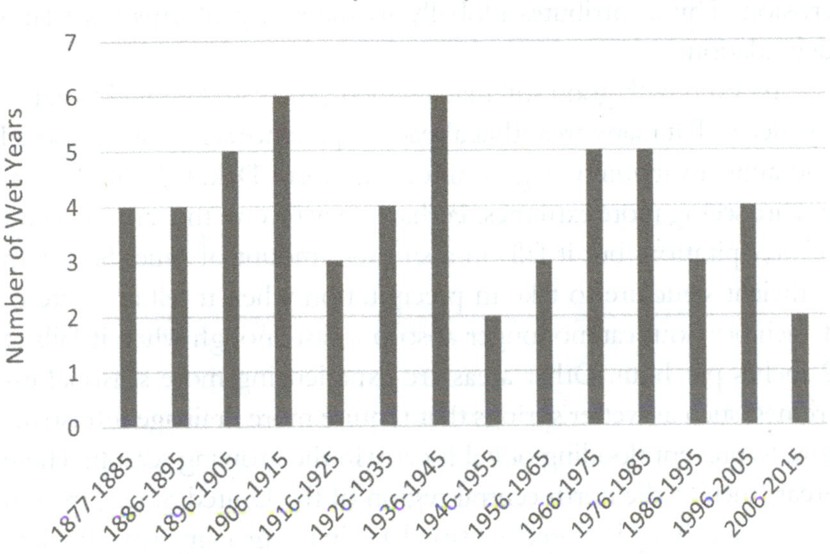

Figure 2.2. Variation in precipitation in Los Angeles over time.

the soil surface without ever entering the soil, because it cannot find a void, is wasted from an agricultural perspective. Why is a void in the soil important? Empty space is a good thing; the voids serve as pathways where water can move into the soil more easily, and where it can be stored for later use by plants. Think of soil structure as the steel beams in a skyscraper. In a soil with good structure, the individual soil particles are held together in aggregates. These larger aggregates cannot become packed as closely together as smaller soil structures, which leaves more void space around them. It is like trying to pack a suitcase; you can cram lots of socks close together, but no matter what you do, there is always going to be a lot of wasted space around the shoes. That space isn't a bad thing when it comes to soil.

A healthy soil has organic matter, a diverse microbial community, and good structure to enhance the infiltration of rainwater into the soil and minimize the generation of overland flow. Tillage and the compaction of soil due to the use of heavy machinery, especially under wet soil conditions, cause a loss of organic matter and topsoil erosion. This contributes globally to massive problems with land degradation.

Yet even with good soil management, increased rainfall variability means that many areas that already experience significant seasonal and annual variability (e.g., California, North Dakota,[9] and Australia) are seeing more extremes. Perhaps they receive the same amount of precipitation, but it falls in a shorter amount of time. Soil with sufficient structure to take in precipitation when it fell at a rate of 1 inch per hour can no longer absorb it fast enough when it falls at 2 inches per hour. Other areas are experiencing more seasonal extremes, such as wetter springs that require more drainage infrastructure to prevent flooding of fields early in the growing season. These areas, such as the north central region of the United States, may in turn be receiving less precipitation later in the growing season when it is needed most by maturing crops. This is leading to the development of methods for storing water drained from fields in the spring so that it will be available for crop growth later in the summer.

This idea of recycling drainage water is an area of growing research in this part of the midwestern Corn Belt. This isn't exactly a new concept, of course. Take, for instance, the rain barrel that Laura installed on a downspout in order to harvest rainwater. The rainwater streams down her roof, collects in the eaves, and drains into the downspout where it pours into the barrel. It is then applied to her flower bed between rain events.

What Laura and other farmers are doing is a variation of the ancient practice of rainwater harvesting. Only the development of mechanisms for how to do it productively in an intensively managed row crop system is new. Rainwater harvesting systems all share the same basic idea. You have an area over which rainwater is collected, a storage vessel, and an application area to make productive use of the stored water.

The famous soil physicist Daniel Hillel advanced the use of rainwater harvesting techniques, such as microbasins, in Israel in the 1950s, based on practices in use by much earlier Byzantine and

Roman settlers.[10] Microbasins are pits dug within the furrows between rows in a planted field, designed to capture overland flow running down the furrow. Rainwater harvesting may include methods of using berms and trenches to direct runoff water to the base of individual plants, or it may represent sand dams built in small, intermittently flowing stream beds in sub-Saharan Africa that fill with water during rain events and store it for community use between storms.

Today, however, various practices designed to capture excess water during the wet season include perhaps a farm pond or better retention in the soil. Growing interest in rainwater harvesting in traditionally water abundant areas like the midwestern Corn Belt reflects a recent trend toward using supplemental irrigation in some areas to buffer risk during dry years.

Green-water resources are the most sustainable, but they aren't the most practical, and, in many locations, they can't meet all of our needs. This is where other irrigation systems, using blue-water resources, come into play. As with all changes to the natural system, however, they come with a price.

Damming the Rivers

Surface water, like green water, is renewable. It relies on the flux of water from the land to the ocean. Rivers run high and lakes fill when wet weather results in rain and snowmelt. When precipitation doesn't fall or too much is being used, the effects are obvious. The lakes and rivers run dry.

Sometimes, however, those lakes and rivers run dry just when you need them to water your crops. This is where irrigation and the construction of large-scale dams come into play. Irrigation has been used for thousands of years, all the way back to ancient Egypt and Mesopotamia. Although the technicalities have changed over the millennia, the basic concept remains the same; getting water to crops where and when they need it. It is a method for delivering water to plants at a regular interval so they can grow strong and healthy and deliver a great harvest at the end of the season.

As alluded to earlier, the story of large-scale irrigated agriculture begins with the heyday of large dam building in the American West in the late 1800s and early 1900s. Building dams in order to create artificial lakes is certainly not new; it was common practice before this time. Think of the thousands of mill ponds still in existence in the eastern United States, a practice brought from Europe.

But we can thank the explorer and geologist John Wesley Powell for the vision that led to the massive reservoirs on the Colorado River—Lake Mead and Lake Powell. Each of these reservoirs can hold multiple times the mean annual streamflow of the entire Colorado River. Powell recognized that, in order to "reclaim" the land of the arid western United States for agriculture, the infrastructure needed would far exceed the capabilities of individual farmers. The government would have to be involved. This period of public investment in irrigation led to complex storage systems along the length of the Colorado River, the Columbia River, and the Missouri River. Other countries have followed suit over subsequent decades.

Although dams may be built for multiple, sometimes conflicting, purposes, including water supply, flood control, recreation, and hydroelectric power generation, the primary purpose of these major western projects was water supply. The purpose of water supply reservoirs is not that different from small-scale rainwater harvesting systems. They are designed to capture the excess river flow in the spring, and hold it until it is needed by irrigators later in the growing season (in the case of smaller reservoirs) or during the next multiyear dry period (in the case of the larger reservoirs).

In this way, dams and reservoirs provide a very real service to regional economic growth and agricultural productivity, but not without other costs. The first cost encountered by most such systems is the displacement of people who live in the area to be flooded by the creation of the new reservoir. Because reservoirs are most often built in remote, mountainous locations with deep, bedrock canyons that can be filled with water, these were often remote, poor populations without much voice in the decision-making process. The flooding of the Tennessee valley to create one of the reservoirs managed by the

Tennessee Valley Authority was the impetus for the character Ulysses Everett McGill to break out of jail in the 2000 movie *O Brother, Where Art Thou?* The McGills' home was about to be flooded, and he only had three days to recover the "treasure" that was left there. More recently, China resettled at least 1.13 million of its own citizens in the process of constructing and filling the Three Gorges Dam on the Yangtze River.[11]

Other costs associated with reservoir construction include the large environmental costs tied with fragmentation of the stream network by the physical blockage of the river system, and the shifts in hydrologic regime from storing and releasing water at times other than the natural cycle. The physical blockage of the river system has become a huge issue in the western United States and Canada due to the cost to migrating salmon populations that can no longer return to headwater streams to spawn. Ecosystems along the Colorado River depended on the influx of sediment from spring floods every few years to create new sandbar habitats. The dams that hold back years' worth of river flows also retain the sediment and yield more consistent flows throughout the year as power generation and the controlled movement of water between reservoirs in the system now supersede the natural rhythm of river flows. In recent years, removal of dams has attracted a lot of attention as a way to mitigate many of these problems.

The Treasure Beneath Our Feet

Our dams control our surface water resources, our rivers. Yet underneath our very feet is a treasure trove of fresh, clean water. These are our groundwater resources, or aquifers. Our aquifers offer up some pretty impressive advantages. But when used improperly, they can be harmed beyond repair, permanently erasing what was once an incredible and necessary source of water. The key lies in the amount of annual recharge. Is enough going back in to make up for what has been taken out? For some aquifers, the time needed to refill an aquifer can be on the order of days or months. But with others, especially those that are heavily used, this could take years or longer, if they ever can completely refill.

It may come as a surprise to learn that shallow groundwater has been used for irrigation for centuries. Water and soil scientist Daniel Hillel has studied the network of shallow wells used to direct groundwater for irrigation in ancient Israel, for instance. Yet large-scale groundwater irrigation was, in most cases, developed after surface water irrigation. As a result, on the global scale, the majority of irrigation comes from surface water resources.

Groundwater irrigation offers the advantage that, to some extent, it is available everywhere with no transportation. An individual farmer can dig a well in the field where he or she needs water. A network of surface irrigation canals isn't necessary to carry water from a distant lake, river, or dam. Upfront costs are usually lower for groundwater irrigation. But the ongoing operating costs tend to be higher. A power source is needed to pump the underground water to the surface. On the other hand, most surface water systems are gravity fed so power is not needed, and in fact power generation may be a side benefit of the reservoir built for irrigation water storage.

In much of the world, groundwater is available for use by whomever digs and maintains the well. Regulations related to surface water usage are far better established, primarily because significant groundwater use has been a relatively recent phenomenon. This means that, in many parts of the world, there is little planning about where and how much water can be withdrawn from aquifers. In regions where groundwater use for irrigation is limited to drought conditions, problems may not be apparent until water is really needed to prevent crop failure. An example of this comes from the drought of 2012 that affected the US Corn Belt. During the drought, a large animal operation in northwestern Indiana began pumping groundwater at higher rates than usual to irrigate its feed crops.

High pumping rates can cause what is called a "cone of depression," a circle of lowered water level around a pumping well. With the drought already lowering regional groundwater levels, this drawdown of the groundwater caused many domestic wells in the vicinity to temporarily dry up. In the Indiana case, dozens of residents in nearby communities were left without water during the hottest and driest parts of the summer.[12] Eventually, the animal operation had to supply the residents with bottled water and pay to dig their wells deeper to provide them with sustainable water supplies in the face of future droughts. During the same period in Missouri, the state agreed to spend more than $18 million to improve more than 3,700 wells with similar problems.[13] These are both regions where groundwater supplies returned to long-term normal levels with the return of wetter weather conditions, but they highlight the need for better management of competition for use of groundwater resources during periods of drought.

More extreme cases also exist in areas that have relied on groundwater for irrigation more heavily and for long periods of time. When more water is removed from an aquifer over an extended period of time (years) than is returned to the aquifer through the slow percolation of rainwater through the overlying soil (recharge), we say that the groundwater is being "mined." The Ogallala aquifer (covering Nebraska and seven other Great Plains states) supplied 17,000 million gallons of water per day for irrigation in the years up to 2000.[14] Although it varies spatially from 0.024 inches per year in Texas to 6 inches per year in Kansas,[15] the average recharge over the aquifer is likely less than 1 inch per year, which works out to just over 8,000 million gallons of water per day.

This mining has led to declines in water levels of over 200 feet and in the total water supply of about 11 trillion cubic feet since 1950.[16,17] Additionally, in some areas, such as in the San Joaquin valley of California, the removal of substantial amounts of water from the aquifer has resulted in a collapse of the aquifer structure as the water historically helped support the weight of the land above it, much like a water bed collapses as it is drained. In such cases, the land surface subsides, reducing elevation, and even if all withdrawals are stopped, the aquifer will never be able to hold as much water as it did before the collapse.

Combined with threats from contamination and climate change, the continued use of large-scale groundwater pumping is unsustainable in many locations and a threat to the long-term viability of regional economies. Parts of the Indian subcontinent are struggling with similar issues, with some locations using water three times faster than the recharge rate. As within the United States, research on India's future water sustainability focuses on better management of existing water resources.[18]

Getting the Jump on Integrated Water Resources Management

It is important to start managing our water resources early in the game. Think of the age-old adage, "an ounce of prevention is worth a pound of cure." Yet, to do so, we have to really understand, and value, the prices we pay by waiting until the last minute. They are, as we have seen, dramatic and damaging. They include the destruction and fragmentation of aquatic habitats, flow disruption, deterioration of water quality, and land subsidence—not to mention the potential loss of the very crops we need for our survival. Although the latter is quite the obvious outcome, much of the former impacts went largely unrecognized, or were considered inconsequential, for a long time. Today, however, things are slowly changing, and we need to keep that momentum rolling.

When water is scarce, we must either develop more water resources or lower the demand for water. In creating more water resources, the first approach is generally to develop new surface and groundwater supplies. Decreasing demand is trickier, and it has been tested in different parts of the world. The problem here is that access to clean freshwater is considered a fundamental right. Putting a price on it (to encourage more careful and decreased use) is often seen as a threat to individual rights. Because of this, globally water is, on average, priced much lower than its true worth. Water rates for water supplied by a public utility cover the cost of water transport, storage, and treatment, but we don't actually pay for the water itself. Some governments have tested out block pricing. In Israel, the government charges more if you use above an allotted amount. This resulted in about a 15 percent decrease in domestic water consumption.[19] In recent years, however, increased efficiencies in the system have allowed the water utilities to lower these highly unpopular tariffs.

When these two strategies stop working (creating more water or lowering the demand), the last strategy is to carefully assess who, among the competing users, gets to use the water resource. In recent years, the natural environment has increasingly been viewed as one of the competing users in the water allocation game. This reflects the growing awareness that there can be real economic consequences due to such things as decreased water quality, wildlife habitat, tourism, and food provisioning when environmental water needs are neglected. In an ironic twist, agriculture has, in fact, emerged as the largest contributor to water pollution in the United States.[20,21] Excess nutrients from fertilizers applied to agricultural fields are transported by the Mississippi River system to the Gulf of Mexico, where they contribute to the hypoxic zone. Hypoxia is a naturally occurring process where decaying algae deplete the oxygen content in water below levels that can support aquatic life, but in the presence of excess nutrients more algae grows, and subsequently areas of hypoxia can grow in size and be sustained long enough to cause significant damage to ecosystems. Over the period from 1985 to 2014, the hypoxic zone in the Gulf of Mexico has averaged 5,300 square miles.[22] Hypoxia has also been a growing problem in over 400 coastal and freshwater ecosystems globally, with the number, size, and duration expanding as a result of anthropogenic contributions of nutrients.[23] The costs of such dead zones are difficult to quantify but usually start with the impact to commercial fisheries. For example, commercial fisheries in the Gulf of Mexico are worth approximately $650 million annually. It is estimated that up to 25 percent of the brown shrimp habitat in the Gulf, an important commercial product, is affected by hypoxia.

Growing awareness of the environmental costs of water resources development has led to a shift in how we approach water resources. It is no longer a strictly engineering point of view, but one that explicitly recognizes the environment. Yet we obviously cannot just do away with agriculture because it pollutes our rivers. We must eat. How to solve this, then?

This problem requires innovative, practical, and integrated solutions. An integrated approach provides all stakeholders a seat at the table, along with hard data to quantify how much water is available and how much is being used. The goal is to explore alternative land use, management, and investment strategies to reallocate water among competing users. Specific tactics may include cost share for investment in more efficient irrigation and appliances to reduce demand; water storage and wastewater reuse capabilities to augment supply; changing crop rotations, retiring land, or developing alternative energy sources to reduce competition; and modifying water policies and treaties to allow flexible solutions.

Keith encountered one such example of integrative management on a trip to Mexico in 2010. While visiting Mexico City, he talked with local engineering faculty who described a planned shift in water use in the basin that supplies the city.[24] This is one of many strategies being evaluated to help resolve the city's impending water crisis. Historically, water that falls in the surrounding mountains was used by farmers for irrigation on the hillslopes. Water then returned to the river system carrying more nutrients, salts, and sediments. Downstream farmers may have made use of it again, but eventually the water not consumed in transpiration reached the water supply reservoirs for the city. Here the water was treated to remove the agricultural residues before being used by city residents, and eventually leaving the local basin and returning to the ocean.

What Keith's colleagues described was an ongoing change in water use, whereby the farmers were being provided gray water from the city as a replacement for using blue water from the river system. Thus water falling in the mountains would drain to the city's reservoirs. The city would use the water, treat it, and then pump it to the farmer who would use it for irrigation. What was not used for transpiration would return to the river system, increasing the overall amount of water in the river system by creating an additional pathway for reuse. This process works because the city's consumptive use is much lower than that of the farmers.

It is not an inexpensive solution, nor is it the only effort required to prevent Mexico City from facing severe water shortages in the coming decades. There are apparently still issues stemming from the lack of communication between various federal agencies and nongovernmental organizations working to alleviate Mexico City's water problems. Mexico City's solution cannot be transferred to Beijing. But it is an example of innovative thinking, which must become more commonplace to assure the world's population of access to water of sufficient cleanliness and quantity, where and when it is needed, to support human and environmental needs.

Chapter 3

The Land That Shapes and Sustains Us

Otto Doering and Ann Sorensen

Our attitudes toward land have changed,
but our reliance upon it has not.

Envision a looming skyscraper in Chicago. There, on the very top floor, resides an exorbitantly wealthy man. He is a billionaire with great affluence and influence. His assets and investments scatter the globe. His penthouse shimmers with gold-trimmed décor, and luxurious carpets cradle the feet of his well-dressed visitors. This man considers himself to be a mover and shaker, someone who influences the fortune and fate of his nation.

Yet he doesn't own any land. He hasn't a single acre to his name. All he actually owns is some square footage of sky. His property is a box of air floating hundreds of feet above Chicago. Indeed, among all of his assets and wealth and investments, not a single piece is soil. Even his companies take up precious little space, just several floors here and there on various skyscrapers around the globe, rented from the owners of the building at a price tidily negotiated by his highly trained staff.

In contrast, imagine a wealthy couple who decide they need a "country retreat" and purchase 40 acres of what used to be prime farmland to build their home. They own their piece of paradise but are totally unaware that they have fragmented the landscape and created problems for the remaining farmers who struggle to move their heavy farm equipment and worry about upsetting the new nonfarm neighbors. Many thousands of other couples happily reside in far-flung subdivisions with names like Kings Farm or Scripps Ranch, named after the farms they subsumed.

Now imagine some beautiful hillsides in Arkansas. In a small, rundown house lives a middle-aged woman on 10 acres of land. She inherited those acres from her father who inherited them from his and so on for generations. She's barely scraping out a living; she panics every year when her property taxes are due, but each year she finds a way to eke out just enough until the next round. There isn't a way to set aside any money. Her savings are next to nil.

Yet, on that land, two of those acres are perfectly tended gardens where she grows vegetables for her local farmer's market. What she can't sell she freezes, and she cans the rest to eat over the winter or to share with her neighbors. In the late spring, local schoolchildren come to visit on field trips. They get to practice planting rows of peas and carrots. She lets the neighbor kids scamper around on the rest of her land, which grows wild. It keeps them out of trouble. They are endlessly entertained by the abundant wildlife on her property.

However, she was approached the other day by a developer who is looking to buy up her land. It is a "long-term" project, the man said, so she wouldn't see construction on the strip mall for at least another four years or so. She doesn't want to sell, but with no prospects and no income beyond what she earns at her pitiful minimum-wage job, what choice does she have? Her guts twist at the thought of her family's land being converted to concrete and not passing on to the next generation of her family, but what is the alternative? She knows what she faces when she retires and can no longer pay her property taxes.

The Land That Shapes Us

When was the last time you sat back and thought about land—what your ownership of it means, whether you even want to own it, and how you might use it if you did? Or how about reflecting on the ways in which other people around you cultivate it? Is it being properly tended, with thought for the generations to come?

Many of us today don't think much about land. As we saw with our billionaire Chicagoan, owning land is no longer a precursor to wealth and influence. And people like this, well-off individuals with good jobs and solid resources, often migrate to big cities where land becomes increasingly "out of sight, out of mind." Another segment of the population heads to the suburbs or further into the countryside, claiming their own piece of land and taking working land out of production with little regard for the consequences. There is a general sense that plenty of land still remains; thus little needs to be done to preserve it for essential uses. Even the essential uses themselves (such as growing food) are only vaguely understood or appreciated as we as a society grow more distant from the land.

Today, only a small part of the population has a direct stake in the land. Our poor, land-owning Arkansan is one such example. Not only does she have a personal and deep attachment to her land, but property like hers is essential for air and water purification, wildlife habitat, recreation, open space, and an environment for quality rural lifestyles and culture. Yet hard-pressed to stay afloat economically, she will soon be forced to succumb to the more lucrative financial rewards of an alternative use for the land of her ancestors. She is only vaguely aware that it is prime farmland, and, although she would prefer it to go into the hands of a farmer, no offer has been made yet, and she doesn't know if one ever will.

This highlights a critical fact. People want land for different things. They want it for housing, transportation, food, fiber, energy, wildlife habitat, to name a few. And despite many of our impressions that our land is vast and plentiful, it simply isn't. There isn't enough land to meet everyone's desires. We don't have the space for everyone's strip mall and everyone's wildlife reserve and all the land needed to grow the food on which we must subsist.

Yet even as its vital importance to us is growing and pressures for its use increase by the day, our awareness of land diminishes. Over time, land has become less and less important as an influence on the American psychic and national identity. We've become more like the man in the penthouse or the couple with their own modest farmette—oblivious as to where our food comes from even as we enjoy its bounty like never before in history.

The United States was shaped by its vast land base, but it has become an afterthought at a time of vital importance. Today, we face threats to the potential of our vast land base to give us what we need to live. By taking it for granted, we put not just ourselves but future generations at risk. Food requires agriculture. Agriculture requires land. And in our hopes to feed the world, the challenge of managing our land is of the utmost importance. In this chapter, we delve deeply into our relationship with our land. As we shall see, it is not quite as simple as it appears on the surface.

Our Evolving Relationship with Land

We can't look at land without also considering how we personally relate to it. There was once a time, many centuries ago, when land "ownership" as we know it wasn't even a concept. Prehistoric humans and many subsequent indigenous communities were hunters and gatherers, harvesting what nature provided with no clear sense of "this is mine" and "this is yours." The earth was just the earth. No one owned it.

This belief has, obviously, changed quite radically. Settlement in America was shaped by landownership. Colonists from Europe wanted land for themselves, and they took it from native populations in the United States through various means. Most colonists did not respect Native Americans' sovereignty, property rights, and titles. If you came to America as a colonist the chances were that you had been born into a tenant family and had never owned land. Landownership meant political rights, wealth, and security.

With so much land available, the settlers fled Europe with its constrained land and lack of opportunities. Settlement in the United States started on the east and headed west as land that was first cultivated lost its fertility and as even more settlers arrived. With the American Revolution, the unsettled lands that had been claimed by the original crown colonies became the property of the new central government. The Louisiana Purchase opened up even more lands, adding 828,000 square miles. The Homestead Act of 1862 granted 160 acres of federal land to settlers for a small filing fee and 5 years of residence. There seemed to be a never-ending abundance of land to go around!

Just take a look at your two authors. One of us owns land that was deeded to a veteran of the War of 1812 and also owns land that was originally deeded to Cornell University under the Land Grant College Act. The other's great grandfather homesteaded in the 1950s in a small valley in northern California. Over generations, the American psyche became one of unlimited land availability for individual private ownership with previously few restrictions, the result being that the sustainability of one's farmland was of less concern because new lands were always available.

Yet this rapid expansion, over time, began to leave its mark. The Dust Bowl is the poster child for the magnitude of shock that could result when rain patterns change, water becomes scarce, and the need to protect soils is ignored. In the Great Plains, farmers converted land by plowing native grasslands. Those native grasses had held the soil in place and retained moisture during dry periods. Throughout the 1930s, these regions experienced severe droughts. The ensuing Dust Bowl resulted in the largest recorded loss of farmland in the United States, extending over southeastern Colorado, southwestern Kansas, the panhandles of Texas and Oklahoma, and northeastern New Mexico. Over 100 million acres were lost to drought and erosion, and millions of acres were taken out of production for decades.

In many areas, over 75 percent of the topsoil was blown away. Some lands never recovered.

During early settlement, there had been little concern with sustainability. The land was treated on the basis of hunting and gathering or something akin to shifting cultivation where one could always move further west for good land. This changed in terms of perception and reality with the Dust Bowl in the 1930s. It led to a greatly expanded role for the federal government in land management and soil conservation.

But memories faded quickly. Many of the same mistakes were repeated during World War II when farmers again plowed up grasslands to plant wheat when grain prices rose. By the late 1970s, deep wells into the region's Ogallala aquifer, intensive irrigation, and giant harvesters were generating immense harvests regardless of whether it even rained at all. Today, the expense of installing conservation practices to minimize the loss of soil sediment and nutrients from all cropland acres that need them is staggering. In Iowa alone, officials calculated that it would cost $1.2 billion annually over 50 years to meet their goals to reduce nutrient and sediment loss from croplands by 45 percent. How much government should be involved in the use of the land is a point of contention, be it enforcing basic good agricultural practices or ensuring sustainability.

We have come a long way over the years. We were once a country of farmers, deeply connected to the earth, with a sense of a never-ending land supply. Yet today, that abundant land supply is gone. And, whereas 200 years ago some 98 percent of the population worked in agriculture, today that number has dwindled to less than 2 percent. At a time when our land pressures are increasing, we have become further disconnected from the land itself.

Challenges

Our land is finite. The uses for said land are many. Our aforementioned Arkansan's 10 acres could become a strip mall, farmland, wildlife reserve, housing, or a stretch of highway. Given her circumstances, it is looking like it will be a strip mall. And although we don't all personally own those 10 acres of land, we reap the benefits and costs of what purpose they fulfill in the end. Even our Chicagoan indirectly experiences the consequences of her decision about whether to sell, and to whom to sell.

We tend to think of farmland as the land that grows our food. Nothing more, nothing less. Yet agricultural landscapes serve many more vital purposes. Not only are they where we grow food, but they are also a source of energy (biofuels) and fiber (cotton to make our clothes). These lands also sustain much of our nation's wildlife. Animals, plants, insects, and mammals occupy these spaces.[1] About half of all threatened and endangered species have at least 80 percent of their habitat on privately owned land and 87 percent is in agricultural production of some type (grazing, crops, timber production).

Additionally, climate change, as seen in chapter 4, plays a large role here. This is a double-edged sword in terms of land; not only do we need agricultural land to grow our food, these very same agricultural soils may be able to sequester about 650 million tons of carbon dioxide every year. This offsets up to 11 percent of US greenhouse gas emissions annually.[2]

And with the increased intensity of rainstorms brought on by climate change, we are already seeing an exponential increase in soil erosion and runoff.[3] It takes roughly 500 years to create 1 inch of soil. Agricultural soils and ground cover, the low-growing, spreading plants that help to stop weeds from growing, help to filter about 70 to 80 percent of the rain and snow that falls in the United States before it enters our streams and rivers. When water enters the soil, it may be mixed with sediment, salts, bacteria, viruses, heavy metals, and so on. The soil itself becomes a filtration system for the water, as though the water were passing through a screen.

In 1991, one of your authors toured Iowa with a delegation of farmers from the Soviet Union. We stopped at a typical, modest livestock and crop farm. It was about 390 acres farmed by a man and his son. A very spirited argument broke out between the translator and the Russian famers. The Russians simply *did not* believe that only two people could farm that many acres. In Russia, it took many times that number to do the job. Needless to say, it was quite the tour.

The development of technology over time in the United States has traditionally been to save labor, not land. One or two people farming vast spaces across America required technological help to get the job done. This drove technological development here in ways it did not in Europe and elsewhere, where land was limited and labor was abundant. Today, with the frontier closed and farmland declining, our new technology is becoming like Europe's. Our labor-saving technologies have been supplemented with land-saving technologies that use intense cultivation, improved genetics, additional chemical inputs, and supplemental irrigation to boost yields per acre. As our prime farmland continues to disappear, it will become harder and harder to coax yields from more marginal lands regardless of how intensively we cultivate them.

This prime farmland is disappearing for many reasons, a primary one being suburban sprawl—the couple determined to have their own farmette. Historically, people tended to settle in these flourishing agricultural areas. As such, these prime farmland areas have also, unsurprisingly, become prime developable land as well. These sprawling developments started back in 1944 when the GI Bill provided money for returning soldiers to educate themselves and buy homes. Suburban areas outside of cities started popping up as competition for land heated up. More people had cars. Interstates were built to connect major cities. Life in the suburbs had become a reality.

By the 1970s, new houses were appearing on large parcels of land. These lands had previously been rural areas. The once dense development structure, of four to six houses per acre at the edge of a city, had evolved into many single homes scattered about on large pieces of land, often 10 acres or more. In the late 1990s, nearly 80 percent of housing built between 1994 and 1997 was outside of urban areas, and 57 percent of new houses were on lots of 10 or more acres.[4] These large lots are too small for traditional farming, ranching, or forestry, so they stopped contributing to any form of rural economy. They led to a loss of open space, loss of wildlife habitat, water quality problems, and a higher demand for county services in rural areas. Today, this trend is manifest all around us.

The Importance, and Challenges, of Measurement and Data

Farmland includes cropland (fields of tomatoes, for instance), rangeland (open country used for grazing or hunting animals), and pasture (land covered with grass for grazing animals). The US Department of Agriculture has documented gradual declines in all three of these over the past years, with grazed forestland decreasing most rapidly. Forestland isn't generally considered to be a part of farmland; however, many farms include small or even quite large expanses of forested lands. There is movement between these different types of farmland. They are changing all the time. Cropland turns to pasture, rangeland is plowed and converted to cropland. Wetlands can even be drained to make farmland, and floodplains next to rivers can be cropped in drier years or protected by dikes.

Given the complexity of the situation, we rely on data. There are two important aspects of this type of data collection. The first is land use. Examining land use answers the question, what is this piece of land used for? What are the objectives or purposes in using this land? For instance, an acre of land on a commercial farm in Iowa may be used to grow corn. The purpose of growing this particular type of corn is to feed animals or make ethanol. It is planted in the late spring and harvested in late summer, and it produces a given number of bushels of corn each year.

Land cover is the second commonly measured characteristic. Land cover is the physical material at the surface of the earth. This can include grass, asphalt, trees, water, and so forth. Today, we can measure land cover using satellite and aerial imagery. It is possible to look at land cover maps over time to observe how things have changed. So this same acre of land in Iowa, in terms of land cover, would be, for a simple classification, cropland. In a nutshell, "land cover" describes the physical land type, and "land use" documents how people are using this land.

Measuring land use and land cover is not as simple or straightforward as it seems. Some of the various US federal agencies, such as the US Forest Service, the Bureau of Land Management, the Natural Resources Conservation Service, the US Census Bureau, and the US Geological Survey all produce estimates for the entire United States. Unfortunately, these different databases don't always agree; they differ in purpose, scope, and how they define various land categories and uses. The data also differ in scale, including extent and spatial resolution, duration, update frequency, and timing.

More to the point, however, what really matters is the underlying quality of the land and its vulnerability. Not all soils are created equal. There are those that can produce crops and pasture plants without deteriorating over a long period of time. Other soil types cannot. In 1961, the US Department of Agriculture developed a classification system. They coded soils using Roman numerals I–VIII: class I soils had few limitations and were at low risk, and class VIII soils were at highest risk. Over time, more classification systems have been developed. If we look at these measures of land quality, we find that our higher-quality land was disproportionately lost in the years between 1982 and 2012. This was often the flat, well-drained land near cities that had been good farmland but was also ideal for subdivisions—a trend we cannot afford to allow to continue.[5]

Teams of researchers are starting to answer the question about how much farmland we need and whether we have enough—at the state, regional, and national levels. They are trying to determine the land requirements for a complete diet, how much food can be grown, and how many people can be fed based on the available land. A good example is California, which has one of the most comprehensive state-level databases on land use. They are beginning to factor land quality and water availability into future scenarios. Today, cutting-edge research uses remotely sensed and mapped databases, data analysis, and predictive models. Additionally, some of the ongoing research right now is attempting to measure the likely impacts of climate change as well as housing density changes. The hope is that good information about where we stand with respect to our farmland communities will provide data to enable states and the federal government to take action and to ensure that we will have enough farmland for future generations, regardless of what might happen.

What Is Enough? And Do We Have It?

Although there *are* researchers who are currently trying to answer questions about how much farmland we need, the last time our government tried to determine whether we were doing enough to protect farm- and ranchland to support future generations was in 1981. That is nearly 40 years ago. That report looked mainly at farmland and our ability to meet short-term world food needs.[6] Even with such a narrow focus, the answer was *no*!

Do we have enough farm- and ranchland to support future generations? We don't. There is a declining agricultural land base and increased pressure for other uses. We will need to act quickly. Until we lay the groundwork for a robust, concerted strategy to secure this resource base for future generations, our nation will be at risk.

The free market by itself will not both preserve and maintain the quality of our agricultural land sufficiently for future generations. It is a matter of basic economics. The value of farmland based on the value of its farm production is usually less than its value to a property developer. The value of this farmland for farming will not increase for agricultural uses unless there are shortages of the products from that land. A free market also won't meet the needs of future generations when current generations prefer doing something else because our children and future generations—those who will depend on our farmland in the future—are not here to participate in today's land markets. Selling high-quality farmland today at the fringe of a city for a high price often looks much more attractive to a farmer than banking on one's children being able to make a better living farming or selling at another time. As explained in greater detail in earlier chapters, there are also "uncertain futures" that a free market won't consider. What will climate change do to productivity? Will our population expand? Will diet adjust to food scarcity?

The free market also does not take account of irreversibility. Some decisions are irreversible for several generations to come. For instance, you can't just flip a switch and convert former farmland, now housing and shopping centers, back into productive farmland again. These changes, for obvious reasons, tend to be permanent. Concrete doesn't just erode back into soil over the span of a few years. The concept of irreversibility is more nuanced still. Farmland is often rotated between rangeland, hayland, pasture, and cropland. This in itself isn't necessarily irreversible. However, converting cropland back to pasture tends to go against farmers' business sensibilities. Why would they want to produce a less valuable crop—hay or pasture fodder—rather than a commodity like corn? It took government intervention to encourage farmers to give up less profitable cropland and convert it to a less intensive use during the Dust Bowl. When farmers focus on short-term gains without an eye toward replenishing and building up their soils, some of the resulting damage can be almost irreversible as well. Once we make these irreversible decisions, we can't just go back on them when we "wise up."

Land also gives us different things, which, as we have seen, aren't necessarily obvious. The Adirondack Forest Preserve in New York helps clean the air—like a giant air filter—as the prevailing winds move on to New England. Yet its original intent was to protect the watershed for New York City after clear-cut logging threatened that water supply. Given the unique and varied importance of different parcels of land we can't just convert any piece of land to farmland willy-nilly. Factors like location, rainfall, water availability, and others must be taken into consideration. This requires active and robust study as well as community engagement and leadership.

Yet it isn't too late, and it is certainly not time to despair. Above and beyond supporting research and wise policy in this area, we can all get involved.[7] Attend meetings of your county planning and zoning board. Question construction activities on prime farmland in unincorporated areas. Challenge large lot developments or far-flung rural subdivisions that cost your community more in local services (like ambulance, policing, and busing to schools) than they pay into county coffers.

You can also support your local farmer's market and help keep farmers in business so that their nearby land remains in farmland. Fight proposed transportation corridors (high-speed rail, new freeways) that don't make sense and just end up generating more undisciplined sprawl and paved-over farmland. Find out if your state has a state farmland protection program that can leverage federal funding to permanently protect farmland threatened by sprawling development. If it does, support the program and the land trusts that buy development rights from farmers eager to keep their land in production for the long haul. If it doesn't, start making noise about the need to develop a program to help conserve your state's land resources. Pay attention to the next federal farm bill (up for renewal in 2018) and advocate for more funding for conservation and for farmland protection.

Finally, take the time to really see our farms and ranches. Make sure your children and grandchildren appreciate this land the way we all should. Whether you are walking, driving, or flying, this is the land that can keep us going regardless of what the future brings. Don't isolate yourself to your high rise, or fail to appreciate your backyard. Our land is precious, and so are the choices we make in how to use it.

Chapter 4

Our Changing Climate

Jeff Dukes and Thomas W. Hertel

Adapting and responding to a new global reality

It's the weekend. You've carved out time in your fast-paced, first-world life to enjoy a quiet breakfast, perhaps with a cup of coffee or tea. There's a light sprinkle of rain pattering softly against your roof. You ponder the day's weather, wondering how the rain will affect your plans. But beyond your immediate 10-mile radius, you probably don't give much thought to the weather or climate elsewhere.

Nonetheless, your breakfast heavily depended on them. Chances are it was a global affair. Tea from one exotic continent with unique microclimates, or coffee from another. Grain from a temperate former grassland plain, sugar from a humid tropical coastal lowland . . . you get the picture. In the developed world, we depend on global supply chains for not just our cell phones and T-shirts but also our food. So even though most of us may not think much about the climate elsewhere (unless, like us, you study it for a living), it matters very much to the food we eat every single day.

Why? Different foods require different temperatures, moisture, growing seasons, sunshine, you name it. Farmers in Canada don't peruse tropical fruit seed catalogues, just as Hawaiians aren't looking to purchase massive quantities of corn seed. The climate of different places determines where each of our breakfast ingredients is grown. And while preparing breakfast in the morning may sometimes feel like a drag, it's nothing compared to how much time it would have taken before the first agricultural revolution that marked our shift from hunting and gathering to settled agriculture. And what played a role in bringing about this revolution to begin with? In part, remarkable climatic stability.

Earth's climate has swung into and out of ice ages since the dawn of time. Earth wobbles slightly back and forth on its axis, swinging from more elliptical to more circular orbits around the sun. It slowly shifts the orientation of its tilt at a given time of year, which affects how much and where energy is absorbed by the planet. These orbital shifts have caused ice ages to come and go approximately every 100,000 years. Over the last 800,000 years, Earth's surface temperatures have swung back and forth by up to about 27°F (15°C).[1] During the last 11,000 years of Earth's history, during which modern civilization developed and agriculture spread around the globe, Earth's position within these cycles locked the climate into a narrow window of variation. This created stable climate conditions ideal for developing and relying upon agriculture, forming the basis for a reliable global food exchange we know today.

Yet in the last 50 years, things have changed. Earth's climate is departing this zone of relative stability. Average global temperatures have increased ever more quickly, with new recorded highs in 2014, then again in 2015, and again in 2016. What's more, the recent warming isn't uniform. High latitudes experience the greatest warming, and large continents such as North America are warming more than islands in the oceans. Why? Well, as most of us know from experience, wearing a black T-shirt on a warm day makes us hotter than a white T-shirt because white reflects energy, and black absorbs it. When Earth warms, snow and ice begin to melt. The energy that would otherwise have been mostly reflected back out to space by these white surfaces instead gets absorbed by darker plants, soil, and water. This causes more warming over the land's surface.

However, it's not just temperature changing. Incidents of rain, snow, sleet, and hail are changing too. These are broadly called precipitation events. As a result, some regions of the world are getting wetter, like the northern United States and Canada, while others are becoming drier, such as the southwestern United States. Also, in many places, precipitation is coming in fewer, more concentrated events, such as the rains that caused flooding in Baton Rouge in the summer of 2016.[2] In the US Midwest, rainfalls categorized as "very heavy events" increased by over a third from 1958 to 2012.[3]

As we'll see across this chapter, Earth's climate is extraordinarily complex. In an effort to better understand Earth's remarkable conditions, scientists create complex climate models, known as Earth system models. These capture our understanding of the planet's atmosphere, oceans, ice, and land. Our models foresee a warmer future everywhere in the world, with particular warming focused over land, as well as across the northern high latitudes. Projections also show changes in precipitation. More rainfall will douse much of northern Europe and some of the relatively wet areas of North America, such as the upper Midwest, while some of the drier areas, like the Mediterranean basin, may become even more parched. Everywhere, rain, snow, and sleet are expected to fall in more heavily concentrated events, with more days in between these events.

What does all this mean to us and our breakfast? Agriculture is one of the industries most exposed and vulnerable to climate change. Our crops, in particular, are extremely sensitive to temperature and precipitation. A late frost in the spring can be devastating, while a heat wave during the critical flowering stage can also result in sharply reduced yields. Excessive rainfall during planting time delays farmers getting into their fields, and flooding later in the season can drown the young plants. Yet too little rainfall during critical stages of crop production is particularly damaging. Additionally, farmers need good conditions to go out and harvest. And even further, the very nutritional content of your food can depend on climate. The concentration of protein, zinc, and iron in your food may well depend on carbon dioxide (CO_2) concentrations in the atmosphere where the crops are grown.

In short, agriculture is the "Goldilocks industry." The weather should not be too hot, or too cold, and rainfall must be "just right." A terrible weather year can mean a poor harvest, or even a total loss of crops. This interrupts the food supply chain. For developed-world consumers, *for now*, this likely means swapping agricultural products from one region to another. Perhaps your coffee comes from Costa Rica instead of Colombia, but your breakfast still looks and costs about the same. Yet, in much of the developing world, where households often depend much more on local products, your breakfast might be entirely changed—in cost, appearance, and even availability.

Our capacity to feed the world depends heavily on climate. And with today's changing climate, it will depend on how we, as humanity, prepare for, respond to, and adapt to these changes.

Climate and Agriculture

Farmers have always adapted to the vagaries of weather. They adjust what they plant, where they plant, when they plant, and how they tend the plants. Do we irrigate there? Yes. Does no-till management work here? No.

This information is usually embedded in tradition and passed down across generations. It is extremely valuable and often makes the difference between failure and success. However, climate change renders this "database" of generational and historical information less valuable. What worked before may no longer apply in an altered climate. The rains come at a different time, or not at all. Spring planting done in late May for generations must now happen earlier to avoid excessive summer heat during critical plant development. The region that was ideal for blueberries now can't grow them at all.

Last summer, we hosted a group of Colombian students and

professors at Purdue University. As a part of that visit, we toured a local farm. One of our Colombian guests asked the farmer about the impact of climate change on his crops. He jumped at the question. Traditionally, they had grown tomatoes on the land we were visiting, he explained. But with increasing heavy rainfall events, the heavier soils in that area became waterlogged, which led to rotten tomatoes. After several years of losing their crops, they had to do something. Their solution? Rent sandier soil on the other side of the county to grow tomatoes. Sandier soil drains more quickly after heavy rainfalls and, as such, the tomatoes now suffer lower disease stress.

Our Indiana farmer then turned the question around to the visitors from Colombia. They shared how the distinction between rainy and dry seasons back home had become blurry. As a result, delicate tropical crops that once set fruit during drier periods were suffering greater stress from disease and spoilage due to higher humidity during fruiting season. Unfortunately for the Colombian farmers, many of their crops are perennial tree or bush crops. They are not easily moved, and moving production across the county would not solve the problem anyway.

The implications of both sets of answers was clear. What you did in the past was not necessarily going to be successful in the future with a more variable climate pattern. And when historical knowledge no longer works, farmers need to rely on other sources of information, such as meteorologists, agronomists, and other scientists. Farmers in the most advanced economies, including the United States, already rely heavily on scientific knowledge, which is often mediated by the private sector or by local extension services.

However, farmers in the poorest countries, in many cases the most exposed to climate change, rarely have access to such knowledge. This doesn't mean that they are unaware of the benefits. During a trip to Tanzania to study the impacts of climate change on their agriculture, we met farmers desperate to obtain medium-term weather forecasts. But with only a handful of weather stations that had been continuously collecting more than a decade's worth of temperature and precipitation data, the necessary information base for such forecasts was simply not available. And without a reliable baseline, understanding potential impacts of climate change was daunting indeed.

Farmers wish to obtain medium-term weather forecasts for very good reason. The weather, and their responses to it, impacts their harvests, or the amount of food a crop provides. This harvest rate, known as yield, is absolutely critical to the farmers because it dictates how much they have to sell, and therefore how much money they can earn. In turn, yield matters to the public, because an abundance of food keeps prices low and ensures that enough food will be available.

One of our collaborators, Professor David Lobell of Stanford University, has closely examined how a warming climate affects crop yields. The speed with which a plant moves through different development stages, including the critical "grain filling" phase, increases with temperature. This means that hotter weather shortens the amount of time a plant fills out its grain, which leads to less harvest at the end of the season.[4] Additionally, plants take in carbon during photosynthesis and release it during respiration. Because a plant moves through its developmental stages faster in higher temperatures, as we saw, the amount of carbon it holds also changes with temperature. As global temperatures rise, the carbon that is lost through respiration increases faster than the amount the plant can take in.

Like us, plants also experience stress. Higher temperatures increase water stress faced by plants. Plants respond to this by partially closing the stomata, which are the leaf openings through which moisture is lost. However, this also reduces the amount of carbon entering the plant, cutting into the plant's growth rate yet again. Extreme heat, particularly during a plant's flowering stage, increases the chances of it becoming sterile. In addition, invasive weeds are often more tolerant of a changing climate. They adapt more quickly and respond more favorably to these changes.

The massive injection of carbon into our atmosphere can also reduce the nutritional value of many crops. Their stalks, leaves, fruits, and seeds develop with lower protein concentrations.[5,6] We've already seen this in rice, wheat, barley, potatoes, and, to a lesser extent, soybeans. This is a concern for poorer populations in developing countries. They, unlike others in wealthier countries, often rely on these staple crops for the bulk of their nutrition. Indeed, one study estimates that two billion people *already* suffer from deficiencies of dietary iron and zinc alone.[7] For these people, climate change–induced reduction in nutrients in their food could have dire consequences.

Yet there are some unexpected benefits. At higher latitudes, frosts at the beginning or end of the growing season can be a problem, and warming can reduce exposure to extreme cold, allowing for a longer growing season. And some of the problems posed by tighter stomata can be offset by the presence of increased carbon concentrations in the atmosphere (oddly enough, these increased concentrations are the major cause of global warming in the first place). For some plants, in some locations, increased carbon levels may even outweigh the negative effects caused by *modest* levels of global warming.

But the benefits of elevated carbon levels will eventually cease, and the damages of higher temperatures will dominate. The question really is *when* this will occur. The Intergovernmental Panel on Climate Change tried to capture this complexity in a report that summarizes the varied impact of climate change over the coming century decade by decade.[8] They found that, although the near-term impacts up to 2030 are ambiguous, as the century progresses, the negative impacts become predominant by far. Indeed, by the end of the century, a fifth of the studies predict harvest losses of over 50 percent.

It is far too soon to give up hope, though. Most of the studies don't integrate a particular human advantage: adaptation. We can change planting dates, plant new varieties of the same crop, change crops, or abandon agriculture altogether and import from a different part of the world. Each adaptation will moderate the impact on human well-being.

However, farmers do not always have an incentive to adopt new varieties. Subsidized crop insurance, for example, can lead to more risky behavior. In the US Corn Belt, farmers have focused on varieties of corn that will generate the highest average yield.[9] This has led to steady growth in average yields, but crop yields in the poorest years have remained largely unchanged. The authors of this study underscore a critical point: a changing climate will throw more extreme events at producers, increasing the number of poor growing years. If farmers are not encouraged to factor this into their decision making, this could pose a threat to food security.

Another roadblock? Adjustments are costly. And so they aren't equally available to all households across the face of the globe. Consider the development of new, heat- and drought-tolerant varieties of maize, which is an important topic on the plant-breeding research agenda these days. The majority of the research into new crop varieties is undertaken in high-income countries. Indeed, the poorest countries in the world account for just 3 percent of global spending on agricultural research.[10] These nations, particularly those in Africa, are likely to be among those most affected by climate change. And without adaptive research and with very limited private sector investment in agricultural knowledge dissemination, many of the developments in wealthier countries will not be transferred to these low-income nations. As our colleague Dr. Uris Baldos discusses in far more depth in chapter 5, technology will be key in helping us to overcome many of the challenges posed by a changing climate.

Civil conflict is, however, another problem. One of the world's foremost international research institutes for dryland agriculture, ICARDA, was located in Aleppo, Syria. Civil war destroyed ICARDA's physical infrastructure and dispersed its talented staff. More generally, it is difficult to retain the best and the brightest researchers in developing countries. Having been educated in North America, Europe, and elsewhere, they find it more attractive to remain in the richer countries where their talents might be more valued and where they feel more productive. All of this adds to the challenge of climate adaptation in the poorest countries of the world, particularly in light of recent evidence that global warming has led to increases in the incidence of civil unrest.[11]

As for the ultimate adaptation to global warming, namely abandoning farming altogether, this is once again a luxury only afforded those in richer countries. This is simply not an option in countries where half of the population works in agriculture. Yet it is in these parts of the world where climate change will be felt most severely in the coming century. The concluding chapter of this book directly addresses what it will take for us, as a global community, to achieve equal access to food. For now, however, we will stick with climate.

The Cost to People and Animals

"The most deadly weather-related disasters aren't necessarily caused by floods, droughts or hurricanes," read a 2015 piece in the *New York Times*. "They can be caused by heat waves, like the sweltering blanket that's taken over 2,500 lives in India in recent weeks."[12]

Crops usually get most of the attention when it comes to climate change, at least from the scientists. But the impacts on the people *doing* the farming are, as the *New York Times* piece so clearly illustrated, truly profound. A series of heat waves in India was responsible for well over two thousand deaths.

Although we are all familiar with the lethargy that sets in on a really hot, humid day, few of us understand how dangerous high levels of heat and humidity truly are. High levels of humidity defeat our human body's natural cooling mechanism, which is the evaporation of sweat from our skin. When the air is very humid, our sweat doesn't evaporate. We can't cool ourselves. So when high humidity combines with high heat, we find ourselves in a very dangerous predicament.

The average US citizen experienced about four dangerously hot and humid days per year during the last two decades of the twentieth century. These were days when the human body couldn't cool itself, and as a result, working outdoors became a major health risk.

By 2030, this level is expected to more than double, rising to nearly 10 days per summer. The situation will be even worse in the tropics, where previous extreme heat waves have already resulted in tens of thousands of deaths. This can precipitate both a humanitarian and an economic crisis. It has recently been estimated that the economic costs associated with heat-related stress in China could rise nearly 200-fold over the next century, reaching annual losses equal to 3 percent of gross domestic product by 2100.[13]

While studies of agricultural workers are rare, people have studied the ideal temperature for working in an office—22°C (72°F). You may also have noticed that this is the setting in your hotel room when you travel. As temperatures fall below or rise above this temperature, productivity falls.[14] At 35°C (95°F), productivity falls by nearly 20 percent. Extended periods of outdoor activity become impossible for even the most physically fit individuals.[15] This isn't as much of a problem in the United States, where farmers can sit in air-conditioned tractor cabs while tending to their crops. However, in much of the developing world, such as India, this luxury is not an option. In hotter, tropical parts of the world, warmer temperatures coupled with high humidity can seriously limit the amount of time farmers can work outside. Intolerable field conditions jeopardize human lives, and they also have the potential to jeopardize much of the world's agricultural production.

Humans are not the only animals affected by high temperatures. Livestock production accounts for about one-third of agricultural gross domestic product and is also vulnerable to climate. At temperatures above 30°C (86°F), farm animals may reduce their feed intake by 3 to 5 percent for each additional 1°C temperature increase.[16] High temperatures also increase disease because conditions for insect breeding become more favorable. These animals have to eat as well, and as such are also critically dependent on crops and grasslands for food. As we saw earlier, crops and grasslands will be heavily impacted, which will in turn also harm livestock production.

It should also be mentioned that our livestock don't respond to temperature in the same ways. Pigs, for example, do not sweat, and they struggle to dissipate heat from their bodies. This is why you see free-range pigs head for the mud on a hot day. Wings can be a distinct advantage for poultry because they allow birds to dissipate heat through the increase in surface area. Of course, these things reverse when it is cold.

Causes

In the 1850s, American Eunice Foote and Irish physicist John Tyndall independently discovered the heat-trapping characteristics of CO_2. Tyndall went on to measure the ability of this and other "greenhouse gases" to absorb infrared radiation, the energy that we feel as heat when we hold our hands up to a campfire or a warm stove. He figured these gases must insulate our planet's warmth from the cold of outer space.

These discoveries came during a time when society was just beginning to unlock a vast wealth of stored carbon for use as an energy source: fossil fuels. As the name "fossil" suggests, these fuels formed millions of years ago. In these long-gone eras, phytoplankton and land plants took up CO_2 from the atmosphere as they do today. However, many of the plants grew in environments that prevented them from returning that carbon to the air—they did not decompose after they died. By trapping large amounts of carbon in ocean sediments and peat swamps, these dead plants thinned the atmosphere's layer of insulation. Over time, much of the peat gradually formed coal, and the richest marine sediments formed oil and natural gas. As people started to burn these fuels, releasing the bound carbon back to the atmosphere as CO_2, the atmosphere's insulation started to thicken again.

Our planet's many varied life forms, including us, depend on greenhouse gases to keep us all from freezing. In our atmosphere these gases allow visible light from the sun to pass through to Earth's surface. Some of those light rays will be absorbed by water, soils, plants, our uncovered skin, and any other exposed surface.

Absorbing this light warms us and our planet. Our planet and our bodies release stored warmth back out in infrared wavelengths. When it is cold outside, we try to retain this energy by wearing layers of clothing, such as scarves, hats, and thick coats. These articles of clothing absorb the infrared radiation and warm up. Our warm clothes then keep us nice and toasty. The greenhouse gases in our air act as Earth's winter coat. They absorb infrared wavelengths year-round, warming up our atmosphere, which in turn makes everything on Earth's surface warmer. Without these gases, Tyndall wrote of England, "The warmth of our fields and gardens would pour itself unrequited into space, and the sun would rise upon an island held fast in the iron grip of frost."[17]

Fast forward 160 years from Tyndall's and Foote's discoveries about greenhouse gases. Society is gushing fossil carbon into the air as never before. The world's growing population and its dependence on energy from coal, oil, and natural gas have led us to produce and release over 35 billion tons of CO_2 per year. By now, the concentration of CO_2 in the atmosphere is 43 percent higher[18] than it was before the Industrial Revolution. This flood of insulation into our atmosphere has warmed the planet, and this warming has come faster by the decade. As of 2017, 16 of the warmest 17 years in Earth's recorded history have all come since 2001, with the most recent years being the hottest of those.[19]

The CO_2 we have released is not doing all the work itself. As the CO_2 we release warms the planet a little bit, a little more water is able to evaporate from Earth's oceans and lakes and the pores of its plants. This additional water vapor in the atmosphere also traps heat, multiplying the effect of the CO_2 itself. As warmer temperatures thaw some of Earth's most carbon-rich permafrost soils, many scientists worry these soils will start a "slow burn" by converting the newly available soil carbon into CO_2 and the more powerful greenhouse gas methane.

Future climate projections make it clear that, over the coming decades and centuries, Earth's climate is in our collective hands. Researchers make these projections by studying how different societal trajectories could affect our future climate. They adopt sets of future scenarios that describe different visions of the future. These societal trajectories consider various factors, including, the rate of technological innovation, how much energy we will use, and the population level, among others.

Currently, the most widely used scenarios in the climate community are known as representative concentration pathways (RCPs). These project future changes in atmospheric CO_2 concentrations. They track higher future emissions rates (e.g., RCP8.5). Lower emissions rates combined with future negative carbon technologies that sequester carbon from the atmosphere result in lower rates of climate change (e.g., RCP2.6). These scenarios make basic assumptions about society that are critical to generating projections for the future. Climate scientists hesitate to use the word "prediction," because the future depends on societal decisions well outside the control of climate science. The differences in how our climate may change in different scenarios are stark, and they truly highlight the degree to which our future climate depends on those very societal decisions.

Under low-emissions scenarios, warming by the end of this century is held to approximately 1.8°F (1°C) over the temperatures experienced around the end of the twentieth century. Higher-emissions scenarios suggest warming by the end of this century to be approximately 6.7°F (3.7°C). The temperature increases over land would be greater under both scenarios, particularly in the higher northern latitudes. The difference in impacts to agriculture, society, and the natural world between the two scenarios is expected to be enormous. As an example, the citizens of Indianapolis, who historically experienced fewer than 20 days a year with a high temperature in excess of 90°F, can expect to experience more than 100 such days a year by the end of the century under the business as usual scenario. However, if the world followed a lower-emissions scenario, this number of hot days could be reduced by over a month per year.

Emission rates will determine how much hotter Indianapolis gets in the future. The following graph shows the difference between the number of days per year that Indianapolis temperatures did or are expected to break 90°F (fig. 4.1). In the graph, the temperatures from 1915 to 2013 are from gridded daily maximum temperatures (when temperatures for a location are deduced from other existing measurement stations) that were observed in Historical Climate Network stations. The future projections come from computer simulations run as part of Phase 5 of the Coupled Model Intercomparison Project (CMIP5),[20] which compared results from 20 climate modeling groups from around the world.

What Can We Do?

Most greenhouse gas emissions come from burning fossil fuels. Coal is mostly burned for electricity, oil for transportation, and natural gas for electricity generation and industrial use (although all of these fuels are consumed for a complex mix of purposes). Agriculture contributes about 2 percent of global fossil fuel consumption.

However, agriculture plays a much larger role in overall greenhouse gas emissions than would be suggested by its fossil fuel consumption alone. CO_2 is released when land is cleared and prepared for new agricultural fields and pastures. Increasing populations of livestock and production of manure generate methane. Fertilizing or irrigating soils stimulates nitrous oxide production. Methane and nitrous oxide, though much less abundant than CO_2, have more potent insulating properties in the atmosphere. In 2010, agricultural emissions contributed more than *one-fifth* of the overall planetary warming from greenhouse gases.

While it contributes disproportionately to emissions, agriculture is also among the sectors hardest hit by climate change. So why not turn to agriculture for potential solutions? The biggest potential lies in the way we use our land. Nearly half of emissions from the agricultural sector come from carbon losses associated with land clearing. Minimizing or stopping future deforestation (and encouraging afforestation) could make a big difference. Not only can we plant additional forests, we can also manage new and existing forests. In the tropics, deforestation has been a massive contributor to greenhouse

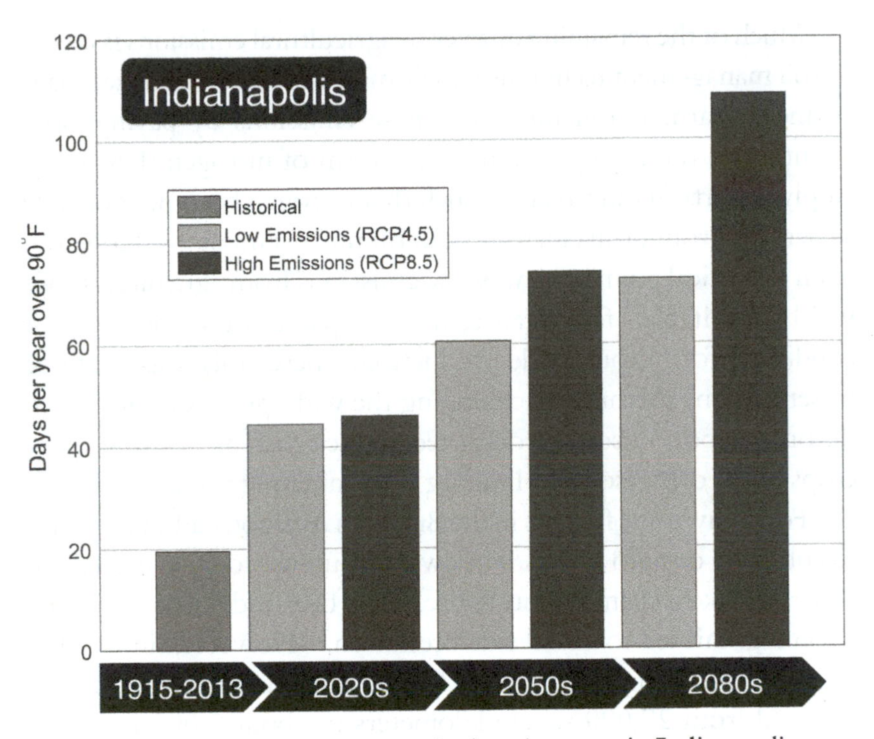

Figure 4.1. Potential scenarios for heat increases in Indianapolis. (Figure courtesy of Scott Robeson and the Indiana Climate Change Impacts Assessment)

gas emissions. Much of this deforestation has been driven by poorly defined property rights, coupled with pressure to expand commercial agriculture. In the first decade of this century, some estimates placed Indonesia as the world's third-largest greenhouse gas emitter due largely to burning tropical forests and draining peat bogs to make way for oil palm plantations.[21] Palm oil exports have been fueled by rapidly growing demand for cooking oil, as well as for biodiesel feedstock.

Another quarter of agricultural sector emissions comes from livestock (particularly cattle). Global dietary choices will strongly influence outputs from the livestock industry, as management of emissions within the industry is unlikely to bring global emissions to a trivial level.

Much of the remaining quarter of agricultural emissions is linked to soil management techniques on farms, particularly fertilizer management. Farmers can minimize these emissions by paying close attention to the timing, amount, and form of nitrogen they apply. Applying fertilizer in a manner such that nitrogen becomes available as demand from plants increases can minimize emissions. However, from a practical point of view, some emissions from agricultural soils will be inevitable—for instance, wet rice paddies provide excellent conditions for nitrous oxide production. These emissions could be offset to some extent by encouraging the widespread capture of additional carbon in soils through techniques such as planting cover crops in the off-season and limiting or eliminating tilling.

For many years, farmers in the Brazilian Amazon had an incentive to cut down or burn forests. They would put livestock on the land in order to stake a claim on future use. With beef and soybean exports booming, this led to massive deforestation.[22] However, since 2004, a remarkable thing happened in Brazil. Annual deforestation rates dropped from 27,000 square kilometers to about 5,000 in 2012–13.[23] This made Brazil the most important contributor worldwide to *reduced* emissions—where the reduction is evaluated compared to what emissions would otherwise have been.

A combination of factors prompted this, beginning with world-class monitoring of deforestation activity using state-of-the-art satellite technology. Where deforestation was found, criminal prosecution was undertaken. Regions where deforestation occurred suffered reduced access to credit and exclusion from the agricultural supply chains. Brazil also expanded the protected areas, introduced payments for preservation of environmental services, and built fewer highways into the virgin Amazon region. All of this occurred *even as* soybean and beef production continued to rise, fueled by more intensive cultivation of existing land area.

As this case so clearly shows, when farmers have the right incentives, they can respond in a sustainable way—increasing production, even as the environment is protected. Unfortunately, in the wake of the recent governance and economic crises in Brazil, deforestation is once again on a sharp rise. This underscores the need for *sustained climate mitigation efforts*. It is not enough to cut emissions for a few years, these cuts must be maintained for the foreseeable future.

One of us (Tom) has tried to estimate what incentive-driven, land-based mitigation policies would mean to the economy if implemented on a global scale.[24] The average farmer in the United States would experience a real income gain of 11 to 15 percent because agricultural prices would rise. In developing countries, this difference is even more pronounced. Average farmers in Brazil would see rising income of nearly 40 percent under some scenarios.[25] Clearly, there is an important economic opportunity for the farm sector if it can engage meaningfully in global greenhouse gas emission mitigation efforts.

Of course, higher food prices aren't good for everyone. The consequences for consumers are striking in the world's poorest countries. The poor suffer disproportionately from the higher food prices that would result from reducing the amount of land available for agricultural production. The message? Any land-based climate mitigation policy must be accompanied by complementary safety net policies aimed to compensate the poor faced by sharply rising food prices.

A Global Problem, a Global Consensus

Ironically, climate change is more a political challenge than a technological one. We already have the technologies in hand to rapidly transition from our carbon-intensive society to a carbon-neutral society.[26] *Convincing* nations to *adopt* these technologies is the hard part. Changing infrastructure requires large up-front investments. Add to this the fact that no single nation alone can solve the problem, prompting a real policy challenge: Why should *one* nation act if others don't?

For many years, climate change seemed far off. Some industries and politicians peddled the message that causes of climate change were uncertain, thereby pushing back the timeline for action. This continues even to this day. And politicians continue to argue about whose responsibility it is to address climate change. How much of the cost should be borne by rich nations? How much credit should go to developing nations for preserving natural resources that regulate our climate by removing carbon from the atmosphere? Some low-lying island nations feel a much greater urgency. Their countries are already shrinking as sea levels rise.

These political issues limited the effectiveness of earlier climate policy actions like the Kyoto Protocol. But political challenges notwithstanding, in 2015 the global community agreed in Paris to work together to limit warming to 3.6°F (2°C). Although the initial commitments made by individual nations aren't sufficient to limit warming to the agreement's temperature goal, negotiators hope commitments will strengthen with time. Political challenges certainly remain, as illustrated by President Trump's decision to withdraw the United States from the agreement. However, other nations remain resolute, including China, the world's top emitter of greenhouse gases. And even within the United States, many states, cities, and top companies are committed to reducing their impacts on Earth's climate.

Climate policy choice boils down to a question of what kind of future we want for our children and grandchildren. If we follow "business as usual," many parts of the world will become nearly uninhabitable without air conditioning by the end of this century. Farm work will grow more and more challenging throughout the rural tropics, where most of the world's poor currently reside. And the natural ecosystems on which humanity depends will be in grave danger.

Yet we *can* still chart an alternative, low-emissions future, provided we do so soon.[27]

Chapter 5

The Technology Ticket

Uris Baldos

History shows us technology is key. Can we innovate, invest in, and accept what we'll need?

Will we be able to feed the world in the coming decades? This is, obviously, the billion-dollar question. Not only do all the authors of this book ponder it in their own personal quest to uncover bits and pieces of the answer, but likely you have entertained it many times as you've flipped through these pages. Of course, we're not the only ones who would like to know the answer. Other researchers, world leaders, policy makers, and concerned residents of Earth are mulling over this tricky question as well.

The question itself, however, is not new. We aren't the first to obsessively and frantically quest for its answer. Many have before, one of the most famous of whom being the Reverend Thomas Malthus, alluded to in chapter 1 on population. Some 225 years ago, this English cleric and scholar completed *An Essay on the Principle of Population*. In it, he argued that food production simply cannot catch up with rising population. He proposed slowing down population growth, or else face worldwide starvation and eventual wars over food shortages.

These are the dire warnings we hear today, are they not? Yet as we learned in chapter 1, population certainly has exploded since Malthus's ominous warning, and we have not faced the dire consequences he predicted. Why? He left out one critical factor in his discussions, a factor that has helped the world avoid this so-called Malthusian trap: *agricultural revolutions*.

The first agricultural revolution, discussed earlier, was the one in which humanity largely transitioned from hunting and gathering to settled communities centered around agriculture. A steady climate facilitated the birth of agriculture, and the birth of agriculture led to a gradual increase in population. But there has not been just a single revolution. As the gears of science cranked and turned, and they have turned very quickly since the 1800s, pressures arose demanding new, sweeping changes to the way we grow and provide food. And so, soon after Malthus's predictions, independent discoveries from plant sciences, chemistry, engineering, and genetics generated innovations and technologies leading to a second agricultural revolution. Farmers could now produce much more than they had before, as farming became mechanized and commercial.

Of course, concerns over world food security did not disappear completely. By the middle of the twentieth century, people were increasingly worried once again. Humankind was recovering from two world wars as well as widespread famine and malnutrition in the developing world. Citizens questioned whether agricultural science could step up to the challenge of feeding the world. With the publication of *Population Bomb* in 1968 and *Limits to Growth* in 1972, the question resurged with great popularity. *Population Bomb* predicted hundreds of millions of people starving by the 1980s. In *Limits to Growth* the authors, wielding a computer model to simulate future food and resource demand, issued a gloomy forecast as well.

Yet again, however, the predictions were off the mark. Between 1990 and 1992, and again between 2014 and 2016, we reduced the number of people experiencing hunger from a billion to 795 million persons.[1] How did we avoid such grim calculations? *Another agricultural revolution*. This third revolution, dubbed the Green Revolution, was a planned, global effort, unlike its predecessor, spanning several private and public organizations. Its intention was to modernize agriculture in the developing world, so as to aid in resolving problems of hunger among the world's most disadvantaged people and later to foster economic growth in the rural areas. It focused on the development of localized high-yielding crop varieties, expansion of irrigation infrastructure, modernization of management techniques, improved access to agricultural markets, and distribution of hybrid seeds, synthetic fertilizers, and pesticides.

The uniting factor that stopped us, as a global community, from fulfilling the dire prophesies preceding both the second and third agricultural revolutions was the same in both cases—technological advances. Technology saved us from spiraling into the mass starvation and war predicted by Malthus in the early 1800s, and from fulfilling dire predictions again in the late 1900s.

Today, we face another well-warranted resurgence of concern. Our modern agriculture is far from perfect, as the many challenges we face so clearly convey. Environmental pollution, unsustainable water use, excessive agricultural land expansion, and more raise doubts over our current food production systems. These beget questions. Can we feed the future without irreversibly damaging nature? Are we up to this challenge? Is it possible to avoid the Malthusian trap, yet again, by pursuing another agricultural revolution—perhaps a "greener" Green Revolution, that will help us grow our food in a more environmentally friendly manner?

By jumping backward in time and working our way forward to the present, we revisit past agricultural revolutions to see how we have managed to transform agriculture with technology each time we appear to be on the brink of disaster. These pages in time give us a glimpse at potential future solutions. Technological innovation can, once more, provide us with the means to overcome insurmountable odds. But what worked before definitely cannot work again. The challenge is whether we can innovate, invest in, and accept the technologies we will need in order to feed the world sustainably.

An Extraordinary Jump Forward

Fertilizers, pesticides, modern plant breeding, machinery, and tractors—we can trace these staples of today's farming all the way back to the Industrial Revolution, which heralded the Second Agricultural Revolution. If we can remember back to our high school history classes, the Industrial Revolution marked advances in engineering and technology that transformed our world. Inventions such as steam engines improved how we transported and produced goods. These and other innovations fueled the rapid growth of the manufacturing sector and encouraged commerce within and across countries. At the time, rising birth rates and better sanitation contributed to a growth in population that provided a much-needed labor force. In times of such widespread change, it is no wonder agriculture was swept along.

So much of today's agriculture, or even backyard gardening, arose from the innovations of this period. Something as simple as the fertilizer my neighbor sprinkles on his backyard begonias, or the packet of "flower food" you pour into the water of the vase holding a lovely bouquet, is a product of this time. Back then, people believed plants could only grow off decayed plant and animal matter. Two men, Justus von Liebig and Carl Sprengel, challenged this mindset.[2] They showed that plants need mineral nutrition. Furthermore, they argued that plant growth is mainly controlled by the scarcest mineral resource, not by the total amount of resources available, otherwise known as the law of the minimum. With better understanding of plant nutrition, the application and production of chemical fertilizers expanded rapidly. Early demand for these minerals was satisfied locally by using available human and animal manure, ash, or bones.

In today's environment of sleek smartphone technology, this isn't the type of technological innovation to which we are accustomed. But two centuries ago, it was a great leap forward indeed.

As agriculture expanded, demand for manure became so large that entrepreneurs found it to be lucrative to import from overseas. As early as the 1840s, Peru and Chile supplied Europe and the United States with millions of tons of guano (colloquially, bird poop). They also exported sodium nitrate. Eventually, available stocks began to decline, prompting major breakthroughs in artificially synthesized fertilizers. Additionally, pest control underwent significant transformation during the earlier half of the twentieth century with the widespread adoption and application of synthetic chemical pesticides.

Jumps forward weren't isolated to just fertilizers and pesticides. Go to the grocery store and look at the fresh produce section. The beautiful fruits and vegetables arranged tidily at your fingertips were artificially bred by selecting desirable plant properties. Since the dawn of agriculture, we were choosing the plants we thought had desirable and useful traits. Back then, however, no one understood that these characteristics were transferred from one generation to the next. As such, success in breeding different varieties of plants with useful traits relied on dumb luck.

Fast forward to the Second Agricultural Revolution, and two seminal manuscripts that were published decades before Mathus's famous essay would have a profound impact on modern plant breeding: Darwin's "The Origin of Species" (1798) and Mendel's "Experiments in Plant Hybridization" (1865). You've probably heard of Darwin's theory of natural selection. At that time, his ideas were quite controversial. He challenged the prevailing belief that species have been created separately. Darwin characterized the mechanisms of evolution, explaining that species with traits well suited to the environment thrive and multiply.

However, Darwin was not able to pinpoint exactly how these traits were passed from one generation to the next. This is where Gregor Mendel's research comes into play. While Darwin developed his ideas by visiting remote regions in South America, Mendel was busy planting thousands of peas and beans in a monastery. He conducted numerous experiments and carefully documented how dominant plant characteristics are typically passed across generations. He also noticed that characteristics from previous generations could reemerge. Later coined as Mendel's laws of inheritance, his work would be at the cornerstone of genetics.

With better understanding of plant breeding, farmers and breeders created hybrids with broader leaves, stronger stalks and roots, and other desirable plant properties. This spurred the proliferation of plant breeding programs for cereal crops (such as corn, wheat, and sorghum and later other food crops) led by both the private and public sectors. The contribution of the public sector through universities and government institutions proved invaluable in the early development and commercial adoption of hybrid varieties. Aside from training breeders, developing and supplying hybrid varieties, publicly funded agricultural colleges and experimental stations educated farmers on the advantages of using hybrid varieties.[3] Lessons learned from the growth of hybrids in the Western world would be reapplied later during the Green Revolution when modern farming technologies were rolled out to the rest of the world.

Yet the advances exceeded even fertilizers, pesticides, and modern plant breeding. Since ancient times, farmers relied on laborers and animals. But there were limitations to using human and animal labor. There is only so much work they can do, in addition to the risk of injury and cost of feeding and taking care of farm animals. So costly, in fact, was the care of farm animals for labor that it was estimated they ate up about one-fifth of the domestic crop output from 1880 to 1920 in the United States.[4] What's more, hiring laborers could be expensive, especially when farm wages were high. This was the case during the rise of the manufacturing sector, which offered higher wages relative to agriculture.

It is no surprise then that when farm machines and tractors were invented and introduced farmers were eager to get their hands on them. Early farm machines made farming more efficient by combining several activities into one process. For example, reapers were pulled by animals and combined both mowing and threshing. This novel machinery attracted the curiosity of farmers and the general public. In fact, when Cyrus McCormick showcased his contraption at the 1851 London World Fair, he won a medal! Then came the self-propelled combine and the tractor.

From artworks to children's toys, the tractor is probably the most iconic and influential farm machine ever invented. While other contraptions made traditional farm labor more efficient, the tractor replaced them entirely. In the United States, it is estimated that tractors alone saved up to 1.7 billion man and animal man-hours in 1944.[5] From less than a thousand in 1910, the number of tractors in the United States increased to 3.4 million by 1950.

The Green Revolution

The third agricultural revolution, also known as the Green Revolution, marked another leap forward for agriculture. The use of the word "green" is confusing today because we have adopted "green" as a common adjective for environmentally friendly practices. With this in mind, "Green Revolution" doesn't imply what we may think. It wasn't a movement centered around the environment. Instead, it was a monumental task aimed at drastically expanding agricultural production by whatever means necessary in countries and regions suffering starvation and undernourishment—even at the expense of irreversibly damaging the environment.

I'm from the Philippines and was born at the tail end of the Green Revolution. This revolution, while perhaps an esoteric or unknown historical term for many, had a profound impact on my life. Without its influence, I daresay I would not have become a professor at Purdue University, and this chapter may very well have been written by somebody else entirely.

Why? Agricultural investments made in Asia in the latter half of the twentieth century were a direct result of the Green Revolution. Both my father and mother belong to a generation of small-scale rice farmers in the Philippines. My father, an agronomist, was one of the numerous Filipino researchers who got to study abroad thanks to scholarships aimed to train professors and experts in order to strengthen local agricultural sciences. Looking back, I have fond memories of snacking on bags of sweet corn my dad brought home from work. Little did I know at the time that he was actually involved in several plant breeding projects on corn and cassava crops, all funded by international agencies. All that time, I was eating corn grown from experimental fields! Those experimental fields and agricultural research institutes in my town, established during the Green Revolution, also strengthened our local economy by creating jobs.

The roots of the Green Revolution can be traced back to World Wars I and II. These wars sorely tested our capacity to feed the world. The devastation was immense. After World War II, Europe, Asia, and North Africa became heavily dependent on the steady flow of food rations and financial aid. To avoid catastrophic famine, governments and international agencies facilitated reconstruction efforts and the restoration of the world's food production capacity.

The solution championed was a modernization of agriculture. In the two decades after the war ended, Europe quickly recovered, owing to its modernized agricultural sector and support of research institutions that were well established even before the war. Immediate transfer of agricultural technologies from North America to Europe, such as maize hybrids, was also possible given the similar cropping patterns and growing conditions. By the late 1950s, Europe, North America, and Oceania supplied most of the world's food needs.

In the rest of the world, however, the amount harvested on any given piece of land, or yield, wasn't increasing much. To grow more, farmers simply expanded cropland, and as a result more forests and lands were turned into fields for planting. Unlike Europe, agricultural research and extension were quite limited or sometimes nonexistent. Most farmers still depended on traditional practices, and they were slow to adopt modern technologies. As a result, food production per person in developing countries either stagnated or declined during the mid-1950s to mid-1960s.[6]

With such slow progress of agriculture in the rest of the world, the global community began to sound the alarms of a looming worldwide food crisis, and this was further fueled by high levels of undernutrition alongside steady population growth. The Malthusian view of the world caught fire yet again. Looking back, we now know that the dire predictions were avoided because the world took unprecedented action to avoid the looming crisis.

Through international cooperation and technological exchanges, agriculture in the developing world was ushered into the modern era. This prompted the development and distribution of local hybrid varieties and modernization of farm production, which transformed hand-to-mouth farming into a profitable agro-industrial sector. Investments in rural infrastructures, such as electrification, irrigation, and road construction, were keys to success. In the case of Mexico, these investments in agricultural output increased yield *almost fourfold* during the period 1940 to 1965.[7] This success paved the way for the establishment of the International Maize and Wheat Improvement Center, one of the world's premier agricultural research centers.

In Asia, the development of hybrids, the offspring of two plants of different species, was crucial to avoiding famine in rice-dependent countries. Adoption was slow, so for some time rice production grew by expansion of cultivated and irrigated areas. However, progress in breeding and adoption of hybrid rice intensified by the late 1960s. The investments in irrigation infrastructure, establishment of the national research and extension system, as well as credit and research programs, quickly incentivized farmers to adopt modern rice hybrids. By 1983, around the time of my birth, around 40 percent of rice planted in Bangladesh, Burma, India, Indonesia, Nepal, Pakistan, Philippines, Sri Lanka, and Thailand was of high-yielding varieties.[8] It is not surprising then that, in the two decades following the rice shortages in 1965, Asian countries more than doubled the national rice output.[9]

For many places in the world, this revolution is more or less finished. But not everyone avoided widespread food shortages. Agriculture in Africa has struggled to keep pace with its population growth. While Latin America and Asia were generally producing staples above self-sufficiency levels as early as the 1980s, several countries in sub-Saharan Africa were experiencing famine and chronic food shortages.[10] There was some successful adoption of modern maize varieties, but these were limited to a few countries and didn't have significant impacts. In fact, it wasn't until the 1990s when Africa's own agricultural research and development programs started to gather momentum.

This delay was a result of several technical challenges. Experts initially thought that wheat and rice hybrids developed in Asia and Latin America could be readily adopted in Africa. But these varieties were designed to thrive in productive soils and in irrigated production systems—both of which were conditions difficult to find in Africa.[11] Unlike Asia, wherein maize, wheat, and rice constitute more than half of the food intake in the region, dietary staples in Africa are far more diverse (e.g., millet, sorghum, cassava, yams, and cowpeas), making it much more challenging to develop and roll out local hybrid varieties in this region.[12] At the start, there was also limited plant breeding knowledge on African staples, further delaying the progress of developing local hybrids. Socioeconomic conditions also inhibited the Green Revolution in Africa, as poor education, political instability, and limited infrastructure prompted one challenge after another. Yet the picture is not all grim—recent evidence shows that changes are being made—however, it is clear that the Green Revolution is still ongoing in much of Africa.

Pitfalls of the Modern Farm

The Green Revolution was hailed as a success by many. Aside from making food more abundant and affordable, it was key to growing the agricultural sector, which in turn helped fuel industrialization and economic development in the rest of the world. Some have even calculated that for each dollar invested in agricultural research during this period around 2 to 17 dollars of economic benefits have been gained.[13] But, in hindsight, the Green Revolution had its downsides and had uneven impacts in society. Although declining food prices were welcomed by urban consumers and kept manufacturing wages from rising, they eroded overall farm revenues and dampened wages of rural labor. In general, poor subsistence farmers were at a disadvantage because they lagged in adopting Green Revolution technologies. They did not have the finances to pay for new hybrid varieties, farm machineries, and agrochemical inputs, so most of the early beneficiaries of Green Revolution technologies were large-scale farmers. Early adopters kept their cost of production low and maintained their profitability despite the observed long-run decline in food prices since the 1950s. As small-scale farming became less profitable, some rural households have been pushed further below the poverty line. Of course, lessons have been learned, and now government agencies typically provide extension and financial support for smaller farm holders so that they can have fair access to improved varieties and modern inputs.[14]

Nature has also been silently enduring the unintended consequences of industrial farming. Agriculture uses a slew of chemical compounds to make sure that crops have sufficient nutrients and are free from unwanted pests and diseases. But these chemicals pollute the environment and often lead to deadly consequences. Let's look at the case of DDT, which was the "silver bullet" insecticide in the past, but which is now banned in most countries. DDT was first used extensively during World War II to prevent insect-borne diseases, such as malaria, from spreading. After the war, DDT became the insecticide of choice in agriculture because little was known about its toxicity. But news of mysterious deaths in wildlife populations led scientists to unveil DDT's harmful impacts as it lingers in land and water, enters the food chain, and causes abnormalities and fatalities in both animals and humans. Now we have switched to relatively less toxic pesticides, but unwanted pollution is always present. Also, nature eventually catches up with our tricks. The Darwinian selection process ensures that surviving pests develop tolerance and ultimately resistance to any new class of chemicals. It is akin to a fast-paced arms race, and if we are not careful in managing resistance, we can create "super-pests."[15]

Overapplication and improper application of fertilizers are also problems. These nutrients are carried off the farm and are deposited in lakes and coastlines, resulting in water pollution. This nutrient-rich runoff from farm fields also encourages growth of algae. Although algae can be an important source of aquatic food, some species are toxic. Also, overgrowth of algae may eventually deplete the oxygen supply underwater after the algae die, thereby suffocating all marine life in a process called hypoxia.

All of that said, we should be careful in blaming only agriculture; households and commercial establishments also contribute to the runoff problem. In 2005, it was estimated that there were 63,000 square miles of lawns in the United States.[16] This is an area the size of Texas and is larger than any single irrigated crop. Most of these lawns receive excess fertilizer that is often supplemented with herbicides and insecticides, whether they are necessary or not. Runoff from lawns goes directly into storm drains and into bodies of water, contributing to hypoxia and chemical pollution of the environment.

This all sounds hopeless, but the next section may lift your spirits.

Toward a Greener Green Revolution

A couple of years ago, I attended a seminar on data-driven farming sponsored by a major tech company. The speakers introduced themselves, and most of them were either programmers or data experts. I was glad to see, however, that one was a young farmer. He was tech savvy and discussed with ease the latest technologies from his farm. Pulling out his smartphone, he demonstrated a range of apps providing him instant access to useful information on farm operations. There was historical crop production data for counties near his farm. Another app reported on the latest crop prices in the futures market. One tracked hourly weather on his farm, a perfect tool for deciding whether to irrigate and when to apply fertilizers. Toward the end of his talk, he readily acknowledged that having this information literally in the palm of his hand enhanced his decision making every day.

The world has never produced so much food as it does today. As our glimpse back in time reveals, this is a result of a series of intentional and sometimes unintentional discoveries and innovations as well as persistent efforts to apply these discoveries and spread them across the world. Yet, aside from the monetary expenses incurred to develop and disseminate the latest agricultural innovations, nature has been silently enduring the unintended consequences of these technologies, and future generations will be faced with them as well. Better crop varieties and machineries improve the profitability of agriculture, leading to further expansion of agricultural lands into natural forests, thus provoking a loss in biodiversity. Excessive application of agrochemicals has led to water pollution and contamination, the effects of which can be seen from space.

Should we ask today's farmers to go back to traditional farming methods that are free from synthetic chemicals and more environmentally friendly? In fact, some farmers have already shifted from conventional to organic farming—using traditional crop varieties and doing away with chemical inputs. The organic farming sector has experienced tremendous growth in the United States and Europe, but its current contribution to world food production is just a drop in the bucket. Globally, the share of agricultural land planted with organic crops is roughly 1 percent, and in general yields from organic farms are around 8 to 25 percent lower than those from conventional farms.[17,18] This means that we need to convert more natural lands into agricultural lands if we pursue this path. Organically grown food commands a premium price in the market, and wealthy consumers can and are willing to pay for it. But how about regular households who spend most of their daily income on food purchases? Completely doing away with modern agriculture in favor of traditional, less intensive farming is not a solution. It would prompt widespread famine and invite a return of the massive food shortages of the 1960s and 1970s. This is not something to which we can return from a humanitarian, political, and social standpoint. Rather, the upcoming challenge is to rethink *how* we produce our food using more sustainable innovations so we can conserve scarce natural resources. Technology provides the answers again as evidenced by the young farmer I met who chooses to use apps and data to better his decision making, reduce application of farm chemicals, and improve his farm's profitability.

For the modern farmer, such as the one described above, a plot of land is no longer just a piece of land. It is a library of information ranging from soil moisture and fertility to weeds and pests. Making the most out of this "big data" is at the center of precision agriculture. Precision agriculture emerged in the mid-1980s when experts began to understand how different growing conditions can be, even within a single field itself. By looking at specific needs with precision, farmers can take a targeted approach to applying farm inputs such as fertilizer and pesticides. This doesn't just reduce costs and wastage, it also limits environmental emissions and reduces risks to human and environmental health.

Precision agriculture runs contrary to current practices in which farm inputs are broadly applied, resulting in environmentally damaging chemical runoffs. Computers, global positioning systems, geographic information systems, as well as sensors[19] all provide the data necessary to give each tiny parcel of the field exactly what it needs. And with the cost of technology going down, most farm equipment in wealthy countries is now outfitted with sensors that can measure just about anything.

Imagine that your tractor can track crop growth, weeds, diseases, and even nitrogen levels and moisture in the soil as you drive around your fields (or possibly as your tractor drives itself). During harvest time, the combines measure grain quality and map yields for each patch of land. All this information can be collected and stored to help in the next planting season, or it can be uploaded automatically via wireless network to software or an app that determines how much pesticides, fertilizer, and water to apply in any part of your field. More recently, drones are being used to collect higher-resolution field maps and to monitor water stress and yield performance.[20]

Despite the huge potential with precision agriculture, it is still quite costly and requires technological infrastructure. In the United States, there is still a long way to go before precision agricultural technologies are fully adopted. As of 2006, yield monitoring systems were applied to around 35 percent of soybean and 42 percent of corn acreages, nationally. When it comes to technologies that allow targeted application of chemical inputs, adoption rates are lower still.[21] Yet, despite these slow adoption rates, we can hope that more farmers will be quick to adopt precision agriculture in order to better manage their resources, lower their operating costs, and improve the sustainability of agricultural production, all at the same time.

Yet we can go much smaller in our measurements with the emergence of nanotechnology. Nanotechnology really drives home the adage that small things can have big impacts. This technology can even further enhance the gains from precision agriculture. Aside from making field sensors smaller and more compact, nanotechnologies can also help improve how fertilizers and pesticides are released. By putting chemical inputs into tiny capsules or in gels, it is possible to control when and how these inputs are released to make them more effective and at the same time reduce chemical emissions and runoff. Nanotechnology also makes it relatively safer and easier to apply and handle these chemicals. There is growing evidence for other useful applications of this technology, such as the nanoparticles in plants that enhance nutrient absorption and plant growth. Some can even be used as insecticides. Nanotechnology is at its early stages and its risks are yet unknown. It might take awhile before the impacts are felt in our food production systems, but it is an area with huge potential to both increase yield and protect our environment.[22,23,24,25]

There is a last technology to mention here. Although extremely controversial in public dialogue, genetically engineered (GE) crops are here—and all indications are that they are here to stay. Since GE's development in 1973, several GE crops have been created and commercialized. For example, crops containing the bacterium *Bacillus thuringiensis* (hence Bt) were developed to prevent crop damage from insects, and they have been adopted worldwide. There are ongoing efforts to roll out GE versions of fruits, oilseeds, as well as root crops. Aside from pest- and herbicide-resistance, plant breeders are also looking to incorporate useful agronomic traits, such as drought- and cold-tolerance, virus resistance, and enhanced nutrient content.[26] Some plant breeding programs offer even more ambitious goals. There is an effort to completely supercharge the photosynthetic process of rice to overcome its current yield limit.[27] Rather than applying nitrogen fertilizer, some plant breeders are looking to incorporate nitrogen fixation in cereals. In the advent of more efficient and precise genetic editing techniques,[28] it is likely that any plans to feed the world will involve the use of GE crops.

Finally, it is important to note that we cannot just rely on new technologies alone to help us solve the problem of feeding the world sustainably. Ultimately these are just tools, and if we do not use our tools properly we may end up repeating history. Information from precision farming will be useless if farmers still choose to overapply fertilizers and pesticides. Although GE crops hold promise, they are not without disadvantages. Like any technological breakthroughs, current GE varieties are becoming less effective as some pests become more resistant. But if combined with proper pest management practices, we can make the most of GE varieties.[29] Not all farmers can afford new technologies, and it is important for governments to foster policies that provide fair access to these innovations.

Innovating, Investing, and Accepting

We face a different set of problems today than we have in prior agricultural revolutions. While a rise in global population is the same, there is also a rise in global incomes driving demand for grains, greens, and meats. At the same time, food production must be balanced with care for our endangered ecosystems and natural habitats. With history as our guide, innovations and technologies are key to ushering in another agricultural revolution. But this time, the revolution must confront the challenge of feeding the world in a *sustainable* manner.

With the abundance of food we have today, combined with a slow (but often irreversible) degradation of the environment, it is easier to ignore the warnings. It is tempting to abstain from action and take things for granted. Yet we must not be complacent. We do not want to risk the well-being of future generations and our planet.

The past has shown us that the metaphorical and literal fruits of agricultural revolutions are often uncertain. They take time to mature and are dependent upon proper planning and sufficient investments in agricultural sciences. Yet, if we hope to discover useful innovations to achieve a more sustainable food system in the near future, we must plan and invest today. What is more, new technology alone is not enough. We will need cooperation and technological exchanges across countries, along with socioeconomic conditions that are both profitable to farmers and affordable for consumers so that the agricultural sector as a whole becomes more environmentally friendly.

Agriculture has changed from the romantic vision often entertained in pop culture, with a horse-pulled plow or a simple red tractor, to something highly mechanized and technologically advanced in many parts of the world. But it must change further still. To protect our environment and grow food sustainably, we must innovate. This demands investment in and support for new farm inputs and crop varieties as well as proper fertilizer and pesticide management. We often wrongly assume that sustainable means going back to the way it was done before. Yet the way it was done before will not feed our world. We can reinvent sustainability through technology with planning and investment, and we have the power and capacity to launch a greener Green Revolution. We simply have to reach out and make it happen.

Chapter 6

Systems

Michael Gunderson, Ariana Torres,
Michael Boehlje, and Rhonda Phillips

The many ways in which food winds its way to our plates

If your drive home from work is similar to ours, then "what's for dinner?" is top of mind. Perhaps like us, you and your spouse, partner, or roommate have the same wrangling conversation nearly every night. It often starts with "dine out?" or "eat at home?" The answer generally depends on what's in the fridge, how much cash is on hand, what we're in the mood for, and how much time we have.

Recently, one of us had some tomatoes stacked on the counter from a haul picked up at the farmer's market. With all those tomatoes at home, BLTs seemed like a good option. But then, we were really hungry, and spaghetti at the local Italian restaurant sounded pretty good too. Or we could swing by our local burger joint and pick up some burgers with all the fixings: lettuce, tomato, onion, and ketchup.

However, if we really were in a hurry to eat, there was always the can of tomato soup in the cupboard. Although perhaps not as quick, but probably more satisfying, was the bag of Italian sausage and rigatoni we had picked up but hadn't found time to try. Of course, there was always the old standby—pizza loaded with marinara sauce, fresh tomatoes, basil, and mozzarella cheese. With all those choices, it's a miracle we ever manage to make a decision.

In this case, though, the BLTs, spaghetti, burger, soup, rigatoni, and pizza all had one thing in common—tomatoes. How did those tomatoes arrive at each of their respective destinations? Did they follow the same path for each and every one of those dinner options? No, they didn't. Dependent upon which meal choice we went for, the lowly tomato traveled a different path from seed in the ground to ingredient in our dinner. And in some cases, a very different path indeed!

It could be from the farmer's market at $3 per pound, or maybe a more frugal route and bought at a discount grocery store for $1 per pound. Perhaps all your tomato consumption comes from Heinz 57 ketchup. Possibly a kind relative or friend shared jars of home-canned tomatoes for use in chili and stews. Could it be that you use Red Gold brand tomatoes that have been aseptically packaged here in our own state of Indiana? Maybe you trust one of the pizza chains to transform your tomatoes for you. For that matter, perhaps you trust the chefs to incorporate tomatoes as they see fit.

Was the last tomato you ate locally grown? Maybe, if you live in Indiana like we do, this is possible for part of the year. But in December, if you want a fresh, locally grown tomato, you might just have to define "local" as anything within hundreds of miles. But there is the chance it was grown with cutting-edge methods like hydroponics or vertical farming in an urban setting. It even could be a FLAVR SAVR® tomato, one of the first genetically engineered crop products to be commercialized. Or not. It could be an heirloom variety that doesn't have that perfectly round shape, but all the flavor you might be nostalgic for. Maybe the farmer sped growth and increased yield with synthetic fertilizers, manure, compost, greenhouse lights, and heat or some combination of these. Or maybe you really don't care because all you can taste is the vodka in your Bloody Mary, and what does it matter what the tomato looked like before it got pureed?

Supply chains are complex systems. Many decisions have to be made all the way from the farm to the consumer's plate. This complexity makes these supply chains difficult to manage, because a change at one point along the pathway has implications for decisions at other places. If the end goal is tomato paste, this requires an array of different steps than a salad tomato, starting with the seed variety and ending with the processing and packaging.

Efforts to improve sustainability and affordability of our food system are so challenging because they must address the complexities inherent to these systems. Our dominant food system today, what we frequently call conventional agriculture, is heavily dependent on efficiencies embedded on a large scale. These are mechanization and agrochemical inputs, such as herbicides, insecticides, and inorganic fertilizers. This system evolved primarily in response to society's demands for affordability and food security over the past several decades. Yet society has now also become increasingly aware that, in chasing this unidimensional goal, there have been and are costs to the environment, the welfare of animals, and other unintended consequences.

Addressing these consequences, and staying nimble and adaptive to new systems, is quite the task indeed. There are a myriad of ways food gets from farm to fork, just as there are numerous ways farmers and ranchers grow and produce your food. Conventional, organic, local, you name it. Yet each method comes with advantages and disadvantages that are often viewed differently by various groups of people. And as such, we face a great challenge. Can we make space for the many ways people get their tomatoes while respecting everyone's preferences, needs, circumstances, *and* the environment?

From Here to There

Many moving pieces and parts play a role in how our food winds its way from start to finish (or our mouths). This path, the supply chain, starts with "inputs," or things such as fertilizers, seeds, and equipment. Regardless of the path, inputs are provided by suppliers and manufacturers, who are kept in business by the farmers who buy their goods. Farmers take these inputs and, with some help from the sun, soil, and weather, transform them into crop and livestock products, such as cattle, poultry, wheat, and corn. These are then sold to processors and handlers who transform them into the food products that you and I buy from grocery stores, at restaurants, or from other food retailers.

Traditionally, the supply chain took standard farm crop and livestock products (frequently referred to as commodities) and smashed/mashed/heated/froze/extruded/pureed/recombined/enzymed/packaged/transported and performed numerous other processes to move them from the farm gate to the retailer or food supplier. In this system, most of the work to produce unique food characteristics was done by businesses past the farm gate, or after the commodities were delivered by the farmer to the local grain elevator, which then shipped them to the processor.

But today, we have increasingly demanded more unique or differentiated food products. Some of these unique features or characteristics have to be developed before the processing stage. For example, organic food or use of appropriate animal treatment and animal welfare practices must be done on the farm. This leads to multiple and often more complex food systems.

When we choose to eat at home, virtually all Americans purchase some, most, or all their food from a grocery store. In US grocery retail, Walmart and Kroger account for more than 26 percent of all the value of groceries sold.[1] The top 10 largest grocery chains sell over half of the food consumed at home. However, there is a wide spectrum of retailers—from "discounters" to high-end outlets, and even our local Walmart or Kroger now offers some organic and local products. Discounters, such as dollar stores selling food at considerably lower costs than other outlets, provide food access to many across the socioeconomic range, although their fresh selections may be limited. Large-scale discounters, such as Costco and Sam's Club, provide discounts in bulk or larger purchases and require membership to gain access.

Then there are providers, such as Trader Joe's, which combines both processed and fresh foods at moderate pricing; small boutiques marketing "gourmet" selections; and even larger chains, including Whole Foods, where many fresh food choices are available at typically higher prices than those of the competition. And here's a twist. Online grocery shopping has emerged, and in a big way. Amazon has entered the business, and there is little reason to assume their entry won't be met with success. This in and of itself could rapidly change the way that many of us obtain our food. Amazon Fresh service offers free delivery for orders over $35, and they propose to offer processed foods as well as local and fresh products. It will indeed be interesting to see how "local" is determined and the impact on farmers, particularly those operating smaller farms in their communities and regions. Walmart also offers delivery service, and, since they are the largest grocery retailer in the nation, it will be equally interesting to see how their delivery services impact the local food supply chain.

Then there is, of course, the option of dining away from home. The fast food restaurant market is, like grocery retail, highly consolidated. McDonald's and Yum! Brands Inc. control more than 25 percent of all food sold in that segment.[2] Yet, in chain restaurants, there is marginally less consolidation at the top end. DineEquity Inc. (with brands Applebee's and IHOP) and Darden Restaurants (with brands such as Olive Garden and LongHorn Steakhouse) control about 13.6 percent of the US market.

As such, on the surface, restaurants don't immediately appear so consolidated. However, they rely on food service distribution companies to source their ingredients. They may pay one business to make available *all* their tomatoes where and when they need them, for instance. The concentration in the food distribution industry is so high that, in 2015, the Federal Trade Commission (FTC) blocked a merger of the two largest companies, US Foods and Sysco. According to the FTC complaint, a combined Sysco/US Foods would account for 75 percent of the national market for food service distribution.

These highly concentrated industries mean that a relatively small set of procurement managers has an outsized influence over our food supply. When McDonald's decides to source chicken that is raised without antibiotics that are important to human medicine, the entire supply chain reacts. Due to the consolidation of larger food retailers, when Walmart decided to sell organic foods, they were able to offer products for 25 percent less than other organic suppliers.

Yet, as we'll see, our new world is slowly reversing this supply chain mentality. Alternative ways of thinking can put the consumer first, and demand chains emerge instead, as there are many ways to gain access to food beyond the typical retail outlets of grocery stores and restaurants. Some of them are more familiar to us, and others are newly emerging as alternative sources. The drive toward improved sustainability will take innovation (chapter 5), communication (chapter 11), and, ultimately, a certain amount of trust between competing interests and opinions.

A Growing Diversity of Demands

We might say we all eat for the same reason, which is to live, right? But beyond that base assumption, different people eat for very different reasons. Some to stave off hunger pains, others for energy to get work done, and some for the nutritional content of their food (too much fat? too many carbs? the right kind of protein? does it have vitamin A?). Most worry about cost, although some can buy whatever irrespective. Others ponder various questions: Does it taste good? Is the tomato hard rather than fleshy? How much time does it take to prepare? Is it contributing to my local community? What are the implications for the environment and my health?

The point? We, as individuals, have very different and diverse demands in terms of the food we eat. And even traditional food consumers, very loosely defined as those of us who just go to the grocery store and buy what we've always bought at the cheapest price we can find, are becoming increasingly demanding in terms of the attributes and characteristics of the products we buy. "Traditional attributes" of food products are what we consider to be nutritional content, taste, texture, affordability, and safety. So, for instance, a tomato has good vitamins, tastes nice, has a good texture, costs less than a buck, and won't make anyone sick. These traditional attributes are the mainstay of most shoppers' expectations, but with time, even the expectations of predictability and reliability of these attributes have increased.

Yet, as we alluded to earlier, the scene is changing. Other factors are emerging to influence what food we buy. Economists call these "credence" attributes. These things can't be directly observed just by looking at food, such as traditional attributes like color, size, shape, and to some extent flavor. Rather, credence attributes are unseen: antibiotic free, certified organic, food miles, carbon footprint, locally grown, animal welfare production practices, sustainable production systems, and so forth. Back again to the tomato—it was grown without synthetic fertilizers and within 100 miles of where it was bought, for instance. We call these credence attributes because shoppers, for the most part, have to take sellers at their word about these things. They are much harder to measure and are often a result of how the product was brought to market along the entire value chain from breeding or genetics to the retail outlet itself.

Because credence attributes can't be seen in the final product, but are processes and activities across the supply chain, documentation and certification rely on tracking and tracing systems. Either a great deal of trust, or data and information systems, are required to monitor and measure these activities at each stage. This information must then be tagged or linked to the physical product so the final product can be credibly marketed and certified. For example, to be able to sell organic certified tomatoes, farmers, processors, and handlers need to be certified by an accredited agency and meet the National Organic Program rules and regulations.

Our food market constantly evolves to meet our changing demands. Take Walmart. As citizens show their increasing concern over credence attributes, such as environmental footprints, even a big box retailer like Walmart is seeking to document things like labor, water, chemical, and other inputs used in the production of goods they sell, because they are aware of the growing credence attribute importance and understand that consumers reign sovereign in the marketplace. Agriculture has also been identified as a major source of greenhouse gas emissions, which has led to greater interest in products produced sustainably as well as sparked interest in reducing food waste and postharvest losses. Farmers are recognizing that commercial agricultural production has its downsides with respect to the environment, and some are adopting technologies to reduce their emissions, such as solar dehydrators, food supply chain systems with shorter distances, cover crops for nutrient management and runoff control, and technologies like LED lighting in greenhouses, or vertical farms.

These efforts take money, though. They aren't free. Retailers, processors, and farmers need to capture increased margins in order to afford these changes. "Margin" is the extra amount of money we will pay for the credence attributes we value. But are we willing to pay for these additional food attributes? They require different and more costly production processes as well as unique and costly tracking and tracing, segregation, storage, and handling, not to mention inventory management processes along the supply chain from producers to consumers. Numerous studies indicate that at least a segment of us are willing to pay some premium for these attributes. But not all of us can, will, or even want to. And there are even those of us who say we will, but when it comes time to put our money where our mouth is, we don't.

Within this context, with growing demands and desires and expectations for traditional and credence attributes, the many ways in which tomatoes can land on our plates are becoming even more complex and diverse. One of the increasingly popular ways is via farmer's markets. Sprouting up across the nation, they increased fourfold from 1995 to 2014, with over 8,650 markets in operation countrywide.

In your authors' relatively small city of West Lafayette, Indiana, we have three farmer's markets weekly where a hot lunch made from locally grown food or a fresh bouquet of flowers can be picked up (not to mention the local honey). They often create a sense of place, connecting locally owned, independent businesses with their consumers. Lending a distinctive "local flavor" to the experience, markets can serve as community resources in some instances. Oftentimes, there is even a festive atmosphere involved with music, food tastings, and other activities to attract shoppers—just ask those who frequent the Crescent City Market in New Orleans! It's normal to pay higher prices at markets of this sort, but some people value the community and entertainment bundled with their food, along with its freshness and being able to meet the source farmer, which makes the prices, to them, a great bargain.

Due to the rising importance of local food systems, the US Department of Agriculture (USDA), through the Agricultural Marketing Service, even created a National Farmers Market Directory. The Local Food Directories website offers a list of additional on-farm markets, community supported agriculture (CSA), and food hubs. The first-ever nationwide survey of farmers selling in local markets was conducted by the USDA in 2015, revealing that over $8.7 billion of foods are sold locally. The number of farmers selling locally increased by 86 percent between 2012 and 2015.

While selling at farmer's markets can be appealing to some farmers, as they can capture price premiums that gain a higher share of the buyer's dollar while farming sustainably, some don't want to or don't have time to sell at these markets. Packaging, delivery, marketing, and selling at farmer's markets means hiring more employees or devoting less time to doing what farmers tend to love most and be best at doing: tilling the soil or caring for animals. Silverthorn Farm, a noncertified 100 percent organic farm located in Rossville,

Indiana, stopped selling at farmer's markets for that very reason, and has focused on their restaurant and CSA customers.

CSAs are direct sales programs that connect farms and producers to a customer base. For those consumers interested in locally grown foods, these programs represent a novel way for creating access to healthy foods. Just as with farmer's markets, CSAs have grown rapidly in the United States, from just a handful to over 1,300 as of a few years ago. Additionally, food hubs have begun to spring up across the country that aggregate locally produced food into a single pickup location for consumers. We have a few here in Indiana—Hoosier Harvest Market, FarmersMarket.com—and others are being developed in different regions to offer local and seasonal food directly from a farmer.

Similar pathways have emerged with farm-to-school programs or to other institutions, such as hospitals. The National Farm-to-School Network reports that over 42,000 schools located in all 50 states have developed such programs, as of 2014. All these direct-to-consumer type programs can provide venues for farms to engage with people, as well as opportunities for community engagement and for consumers to learn more about food, including where it is grown and how. Although markets are limited for most producers via these pathways, it can help stabilize a base of customers for farmers in a community or region.

There are, as we have seen, many pathways to getting our food, whether we're interested in mostly eating out, cooking in, or connecting directly with farmers. But whatever our choice, the future is bound to unfold with even more options.

But Where It Mostly Comes From . . .

We spent the past couple of pages exploring all these unique ways in which we can get our food. Farmer's markets, CSAs, organic—they are interesting indeed, and at times carry certain advantages. But they don't feed our entire population by a long shot. In fact, most of our food comes from one single system, as we alluded to earlier; the one we call conventional agriculture or, alternatively, production agriculture.

The roots of production agriculture in the United States grew out of the past. Farmers became more and more efficient in growing food over time, and they continued to build on those efficiencies as a way to deliver on consumers' desires for lower prices and increased availability. Back in 1840, our population was much smaller than what it is today. Some 17 million people lived in the United States and, of those, about 9 million were farmers. Farmers made up just over two- thirds of the working population itself. Now if we jump ahead to 1910, the United States had grown to a population of 92 million. Of that, there were 32 million farmers, making up about one-third of the labor force. Today, however, with a population of 314 million, there are only 2.2 million farmers. They make up a mere 1.5 percent of the labor force.

Why did the ratio of farmers to other Americans become so small? The answer is conventional agriculture. If our food system was what it was back in 1840, about 110 million people would have to be directly involved in farming. But instead, there was a major advance in how much any one single farmer could grow, which freed up almost 108 million working Americans to do all kinds of other things. Because we were no longer so tied to agriculture, numerous other options emerged for people to explore when they were seeking a profession.

Today, conventional agriculture meets our broad desire for safe, inexpensive food in North America. It delivers the cheapest, most secure, and most abundant food supply in the history of humankind. It has resulted in technological innovation, which means we grow more than twice as much food now relative to 1950, using 2 percent fewer resources (e.g., land, chemicals, fertilizers, and equipment). A US farmer fed about 19 people in 1940; today a US farmer feeds about 155 people (fig. 6.1).

What, exactly, is conventional agriculture? Broadly speaking, it is a category of farming that employs certain practices to produce

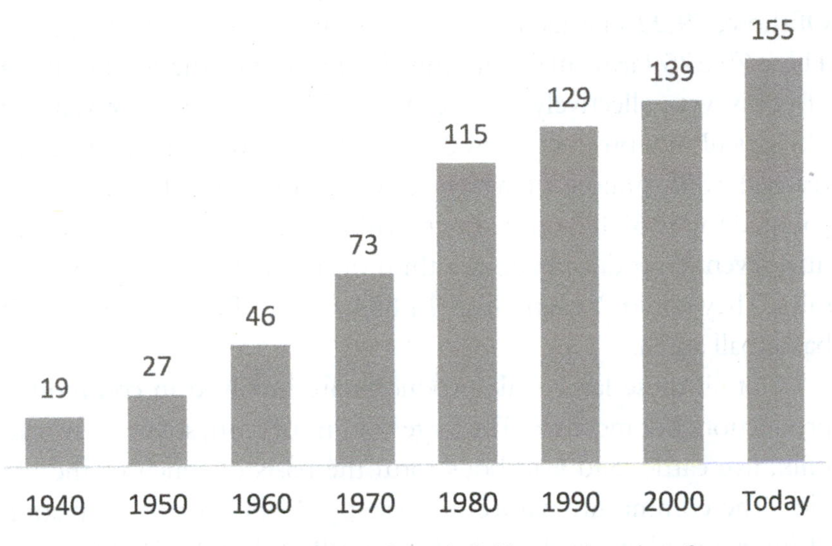

Figure 6.1. Average number of people fed by one American farmer.

the greatest amount of food possible. This can mean using synthetic chemicals or other inputs. It prioritizes food security and economy while also incorporating technology to minimize the impact on the environment. Think of great, expansive fields of a single crop, where the farmer uses the most efficient techniques at hand to grow as much food as is profitable per square foot of land.

Traditionally, the US food system has focused on enormous production of commodity agricultural products, such as corn, soy, and wheat. For instance, each year nearly 90 million acres of fields are dedicated to growing the exact same kernel of corn, known on the Chicago Board of Trade as #2 Yellow. Each producer grows exactly the same product as all the others, which means buyers can choose the producer offering corn at the lowest price.

Today, farms using conventional growing methods tend to raise the most amount of food. To get a better understanding of this, we can look at some data. The USDA defines a farm as, "any place from which $1,000 or more of agricultural products were produced and sold, or normally would have been sold, during the census year." According to the USDA's 2012 Census, there are 2.1 million farms.

Of those, 79,225 farms had annual revenue in excess of $1 million. These 79,225 farms make up only 3.8 percent of the total number of farms, yet collectively produced two-thirds of the entire value of US agricultural production in 2012. Total agricultural sales in 2012 reached $395 billion, of which fewer than 4 percent of farms accounted for $262 billion. Farmers with more than $5 million in annual revenue produced about a third of the value of all agricultural sales. They couldn't even fill all 14,804 seats in Purdue University's basketball arena.

Not all these large-scale producers are involved in commodity production, but most are. The large volume of corn, soybean, wheat, milk, live cattle, and lean hogs form the basis of much of the US diet. The commodity nature of these products led to the creation of futures markets, such as the Chicago Board of Trade. The sheer scale of production is mind numbing. There are more than 14 billion bushels of corn, nearly 4 billion bushels of soybeans, 1.5 billion bushels of wheat, 210 billion pounds of milk, 25 billion pounds of beef, 23 billion pounds of pork, and 38 billion pounds of chicken produced in the United States annually. Because a bushel of corn weighs 56 pounds, the total production weighs about 392 million tons. Coincidentally, that is remarkably close to the estimated weight of the global population, which is 316 million tons.[3]

Conventional agriculture, while not without costs, has allowed us much as a society. It has freed up large swaths of the population to accomplish goals apart from meeting our caloric needs. Farm jobs are often less physically taxing. When it comes to hay, for instance, most hay is cut, baled, transported, and stored using mechanized equipment. It is even possible that a human hand would never touch alfalfa from the time it is planted in the ground to the time it becomes a part of a cow's cud. This matters, because this type of labor can be dangerous. Among the Bureau of Labor Statistics occupations with the greatest risk of fatalities are loggers, fishers, farmers, ranchers, and agricultural managers. That said, we have managed to substantially decrease the amount of risks inherent in working long hours with machinery and animals.

What is more, we spend less of our income on food than we used to. Total food expenditures fell from just about 18 percent of personal income in 1960 to less than 10 percent by 2014. If the average American earns about $54,000 of income, the average American would have about $4,000 more to spend on goods and services other than food today compared to 1960. Over a lifetime, that savings could represent another house! The reduction in food's share of total expenditures would be even larger if we weren't ramping up our food expenditures away from home at restaurants, where we are really buying food preparation and cleanup services as well as good things to eat.

Even as prices decline, our food gets safer. Dr. Jesse Wagstaff, a toxicologist with the Food and Drug Administration, has suggested that food safety regulations save 1.8 million deaths *every year* compared to what would have occurred if food safety rules had remained unchanged since the initial passage in 1906 of the first Pure Food and Drug Act and the Meat Inspection Act. Food safety is a priority of the conventional agriculture system, one which we often take for granted, as we so rarely fall ill from food anymore.

Complex Systems, Complex Costs and Benefits

Yet, in recent decades, agricultural technologies have raised a series of concerns and critiques. For example, we now understand that agricultural runoff from the Mississippi River basin contributes substantially to a dead zone in the Gulf of Mexico. Pesticide resistance in weeds and insects has arisen through expedient short-term activities. Methane emissions from confined livestock and poultry are significant contributors to climate change.

These criticisms have arisen as the scientific community's understanding of these complex natural and biological systems, as well as consumers' awareness, have increased. Income growth in the United States is a large factor at work in this renewed interest in how our food is produced and processed. We spend a far smaller share of our total income on food than our grandparents did. This gives us more latitude to care and more ability to pay for sustainable food. It is fueling a revolution in the development and application of sustainable practices. This is also fueling economic opportunities for farmers, processors, retailers, and the sellers of technologies, as we are at an important crossroad where the demand for sustainable improvements is outpacing the scientific community's base of knowledge about the consequences of change within our complex food systems.

Yet the complexity of solving these issues is further complicated by the biological nature of food production. Making sustainable changes can take a long time, and this accumulates costs to farmers, often before they can realize a return for their efforts. Cover crops (crops planted in the off season to prevent erosion and nitrate leakage and promote soil organic carbon) are one such example. The benefits can take years to accrue. Understanding the economics of such long-term biological processes is complicated and not well researched to date. Additionally, the complexity of the systems is such that one positive change can lead to new challenges. Poultry cage housing was eliminated in California, which changed the view of the poultry industry nationwide, and key retailers like McDonald's now favor egg suppliers who use alternative housing systems like enriched colonies or aviary houses. Aviary houses are the most like free range, in the sense that the chickens are still in an enclosed environment protected from weather, predators, and disease-carrying migratory birds, but they have an extensive area covered in soil or litter (wood shavings, etc.) in which to perform their natural scratching and pecking activities. Their nesting areas are elevated to imitate the natural nesting sites of a bird. This sounds great from a chicken's perspective. Research has uncovered, however, that a lot of chickens scratching in the litter creates dust that is hazardous to workers and perhaps even the health of the birds.[4] These are probably not insurmountable challenges on the way toward sustainability, but they illustrate the complexities we face with respect to the design of our food system.

Antibiotic resistance is another good example. For years, farmers had over-the-counter access to antibiotics that are medically important to humans. They were often used therapeutically to enhance growth of livestock or poultry. We now understand that this contributes to antibiotic-resistant strains of infectious bacteria. However, according to a 2017 study by the Pew Charitable Trusts, the sale of antibiotics that are medically important to humans for agricultural use increased by 26 percent between 2009 and 2015.

This highlights another complexity in our market system—that of free ridership. When we buy our food, the price we pay seldom reflects the external costs to the environment or humankind that may be incurred in the production and processing. We faced the same complex problem with air and water quality in the United States in the 1970s when manufacturers, energy producers, and car drivers did not face the costs associated with environmental and human health degradation resulting from their actions. The remedy was to regulate their emissions technology, create markets for their emissions, and tax behaviors or key inputs like gasoline. The good news for antibiotic use in agriculture is that, starting in January 2017, all human medically important antibiotics used in agriculture essentially now require the equivalent of a prescription from a veterinarian. We should keep our eye on the data collected by the Pew Charitable Trusts to see what happens. We should also probably expect our meat, milk, and eggs to cost a bit more until farm managers find new ways to manage disease risks.[5]

The increased awareness and concern about how our food is produced and processed has raised demand for locally grown, low-processed, chemical-free, and pesticide-free foods that are also free of genetically modified organisms (GMOs). As a result, the credence attributes we discussed earlier have emerged. Additionally, more Americans are supporting and buying from small farmers selling in local markets, such as the farmer's markets alluded to previously.

However, we also lack complete understanding of how sustainable these various food systems are themselves. On the surface, organic sounds very sustainable, but weed control in such systems often involves dragging mechanical cultivators through the field to root out weeds. They also disturb the soil, leading to organic matter decay and the release of carbon dioxide, a greenhouse gas, or the use of weed flamers to burn off weeds between crop rows, which also emit carbon dioxide.

The tricky piece of all this is determining how to move forward, and how to balance the costs and benefits. For instance, while local markets are popular, not everyone is willing or able to pay for locally grown foods. Local foods can be more expensive and sometimes unaffordable for many. There are also, however, an increasing number of Americans who want to grow their own food. Urban and suburban families are spending more money in growing herbs, fruits, and vegetables than in previous years. Among these, millennials, baby boomers, college graduates, married households, and families with income of $75,000 or higher are gardening the most. These demographic groups are spending over $3.6 billion per year on plants and seeds to grow foods at home. Yet not everyone has the time, or the land, to do so. And so, some communities are engaging families in urban, suburban, and rural areas to grow their own fresh produce. Americans growing their own food are reported to have a higher sense of integrity and validity about the food they consume. They like food that has a shorter supply chain, which means that it takes less time and resources for food to travel from the farm to the fork.

The giant retailer Walmart, as part of their sustainability program, pledged to support small businesses by more than doubling the fruits and vegetables purchases from local farmers, going from 4 percent of purchases in 2010 to 11 percent. However, local produce for most large retailers is supplied from large-scale outdoor farms with annual sales in the nine digits. Smaller farmers, on the other end, are typically unable to capture this highly profitable market due to volume, price, and delivery requirements. Controlled-environment agriculture (CEA), one of the trendiest movements of the local food system, is capturing part of this demand of large-scale local food production by growing food in indoor setups and closer to big cities. CEA using hydroponic, aeroponic, and aquaponic systems represents only a few examples of the futuristic twist evolving in agriculture to create indoor farming and supply large amounts of food products to markets located in larger populated areas.

CEA systems aim to provide optimal growing conditions by converting unused buildings or containers into a combination of plant science, computer-managed technologies, and engineering. CEA optimizes the use of water and space to grow from 10 to 20 times more pounds of crops in the same land as outdoor farms. A CEA farm can use warehouses to produce almost a million pounds of leafy greens, basil, and mint per year. Hydroponic and aeroponic systems replace soil substrates with water or nutrient solutions to grow plants. Hydroponic systems have been especially popular among larger-size commercial operations or operations adopting the USDA organic certification.

Newark, New Jersey, is the home of the world's largest aeroponic farm: AeroFarms. This operation, built in an economically depressed area, hopes to bring jobs and two million pounds of leafy greens to one of the largest population centers in the United States. AeroFarms estimates that it uses 95 percent less water than what the same crops would use outdoors, while producing 130 times more leafy greens per square foot. Water recirculation, chemical-free growing, low transportation emissions, and growing crops in vertical stacks are key components of the AeroFarms controlled growing technologies.

FarmedHere, a suburban Chicago 90,000-square-foot indoor organically certified aquaponics operation, started in July 2015, is growing leafy greens in soil-free environments. Plants were fed with a nutrient-rich water solution and maintained in an herbicide- and pesticide-free environment. The plant nutrients were recycled from the adjacent tanks where hormone-free tilapia grew. Asserted advantages of such large-scale urban indoor farms include the proximity to their clients year-round. The farm systems use supplemental lighting and heating, while offering potential efficiency increases in the use of water and nutrients. FarmedHere vowed to achieve these asserted advantages and attracted much media attention; however, they had to shut down their operations in January 2017. FarmedHere CEO Nate Laurell said profitability was the main reason for their decision.

FarmedHere is not the only vertical farming project being abandoned. While the profit potential of indoor CEA farms can be high, these operations are capital-intensive, startup costs are high, and farmers and investors face much risk and market uncertainty. Many of these high-tech farming operations are growing crops vertically by adopting the latest technology in the agricultural industry: LED grow lights, computer-integrated monitoring systems, climate-controlled grow rooms, and considerable electricity consumption that can only be offset by growing high-valued, high-profit margin crops. At the same time, more frugal attempts at extending the growing season for vegetable crops and bringing production into closer proximity to urban consumers are growing. Low-investment, high-tunnel grow houses with minimal lighting and heat cannot produce year-round, but they can add weeks or months to the normal growing season in both spring and fall.

Urban farming, in the form of community gardens, is a fast-growing form of agriculture that possesses unique benefits, opportunities, and challenges. Urban farming can help improve access to healthy and fresh local produce while creating economic opportunities, providing a source of labor, reducing transportation emissions, and fostering the social fabric of communities. Community gardens have become popular because of the social benefits they provide to communities, such as neighborhood beautification, community networking, and use of vacant or underused lots for garden development. Community gardening is also popular among churches, schools, nonprofit organizations, social service agencies, and food banks. These community gardens are improving neighborhood appearance and providing residents with access to fruits and vegetables, while also contributing to youth training and bettering residents' health.

The USDA recently announced the availability of up to $8.6 million in grants to "fund community food projects to enable low-income communities to address food insecurity and hunger by strengthening local food systems." These food initiatives aim to improve the access to technologies to smaller farm operations, contribute to the development of communities, and increase biodiversity among farming operations. However, these urban systems can be more expensive than traditional agriculture. Space in urban areas is limited, and waste from urban agriculture can increasingly become a problem as more farmers grow fruits and vegetables in the inner cities.

Looking Ahead

Trends point toward an increasingly complex value chain, starting with the very first production inputs—seed and genetics. Shoppers will increasingly push for more information in addition to its geographic origin. On the production side, technology is increasing the likelihood that the supply chain can actually offer this detail in a cost-effective manner. Technology has increased the precision of farming down to field units smaller than acres. Scanner and RFID technology will enable low-cost tracking of food as it moves through the supply chain. Management information systems will improve communications among all links in the value chain.

Maybe, in the future, when we pick up a tomato, it will have a QR code lightly etched into its skin. A scan with a smart phone will quickly provide information relevant to the consumer. An app may allow us to prioritize the information important to us, by cutting quickly through the mountains of information known about this one tomato: weight, calories, variety, geographic origin, farmer's name, picker's name, miles traveled, carbon footprint, production method used, fertilizer used, irrigation system used, transportation method used, food safety measures taken, packaging material, and more.

Those of us in wealthier nations are already becoming increasingly responsive to how and where our food is grown as we work toward a more sustainable system. For those of us with the income and willingness to spend extra, we can afford to get what we want and eat those costs, so to speak. Many of the specific characteristics we demand are supplied to us. Great!

But not everyone can afford to do so. Therein lies the danger of insisting that everyone adhere to our food system of choice. Not everyone can, or even wants, to eat organic. Not everyone can, or even wants, to pay extra at their local farmer's market. Not everyone wants to eat conventionally grown food, which is also fine. While we may feel that one system is superior to another, we can't assume that others agree. Nor should they have to. What is right for us and our circumstances is often not viable or palatable to others who live under different circumstances.

What is important is that we come to agreement on what we need for the sustainability of our food systems. Once we have built this broad consensus, then we can use it as a guide to help make personal and public policy decisions. Consider, then, a myriad of different but parallel systems that attend the needs, preferences, and ability to pay of our diverse society. For this, we can rally behind a diverse approach, which allows for many types of systems meeting many types of needs and desires.

Chapter 7

Tangled Trade

Thomas W. Hertel

International trade creates an invaluable buffer against unforeseen food shortages. Yet can we overcome the tangled trade barriers blocking our path?

My interest in international trade was sparked over 30 years ago. In the early 1980s, representatives from nations around the world gathered in Geneva, Switzerland, to try to launch negotiations around a global trade agreement, but the conference stalled on agriculture. Nothing came of it, and it was widely regarded as a failure.

The failure itself was not what garnered my interest, however. It was what came next. Four years later, something remarkable was born out of those unsuccessful negotiations. Countries from around the world launched a new round. This one, the Uruguay Round, so called for having started in Punta del Este, Uruguay, in September 1986, was to be the second attempt at finding common ground. Although twice as long as intended, it yielded fruit. Over the span of seven and a half years, 123 countries came to agreement around the highly contentious and troublesome matter of international trade.

The result? The establishment of the World Trade Organization (WTO) on January 1, 1995.

For me, an agricultural economist, this was simply extraordinary. Before the WTO, there was the General Agreement on Tariffs and Trade. It had sharply lowered tariffs on manufactured goods, causing trade in these products to grow by leaps and bounds. But trade in food and agricultural products was a different story. A complex system of quotas and variable tariffs caused trade growth in this area to be sluggish. One of the major goals of the Uruguay Round was to smooth the way for increased trade in agricultural products so they might achieve more of the success obtained for manufactured goods.

Cross-sector trade-offs were critically important to the success of the Uruguay Round. For instance, if Japan was going to lower protection for agriculture, it needed to be shown there were gains to be had elsewhere, such as in manufacturing exports. When it came to the United States, dramatic losses were expected in the textile and apparel sectors. Would there be gains in other sectors to make up for these losses? Before buying into the agreement, national negotiators wanted to be assured that they would obtain improved market access for their country's exports.

Here's where I came in. Trying to measure, or quantify, these economy-wide gains became a key part of my early research. Ultimately, it led me to create an entirely new center at Purdue University dedicated to measuring the impact of international trade agreements on imports, exports, employment, and economic welfare, called the Global Trade Analysis Project. Based on the results of our data and research, other economists and I concluded that, for most countries, the expected gains from the Uruguay Round would be large enough to offset the likely losses, and so "the gainers could compensate the losers."

GTAP

This compensation between gainers and losers rarely happened in practice, though. This was the cause of much of the backlash against these trade agreements. This is a critical topic to which we will return later.

- effect for trade agreements

The creation of the WTO further cemented "rules of the road" for international trade. Since 1995, many other countries have joined, including China in 2001. This common set of rules governing international commerce has facilitated an explosion in trade. The shirts we are wearing, the cell phones on which we are talking, and the cars we are driving were often assembled elsewhere, using components produced on multiple continents. Globally, we have become increasingly dependent on other nations in our daily lives for both work and play. Companies, likewise, think increasingly globally when it comes to developing and marketing products.

This increased interdependence has improved our standard of living. While not deterring regional wars, it has contributed to an extraordinarily long period free of world wars. Indeed, this was a key part of the case for free trade made by Cordell Hull, known as the father of the United Nations. He argued that countries trading with one another would be less inclined to go to war. In 1945, he was recognized with the Nobel Peace Prize. Hull's admonitions are well worth keeping in mind. International trade's importance extends well beyond the usual economic metrics such as gross domestic product and employment.

When it comes to our food, today international trade is fundamental. Unlike manufactured products or banking services, agriculture is inextricably intertwined with climate and soil. These, as we have seen in earlier chapters, vary widely across Earth's surface. Some areas are well suited to farming, some are not. Furthermore, regions of the world with defined seasons, such as the US Midwest, can't grow food outside of greenhouses in the cold winter months. The only way people who live in these regions can enjoy fresh fruits and vegetables throughout the winter months is to import them from the tropics or opposite hemisphere. As a result, even in Indiana we can have fresh blueberries with our cereal in January, something unheard of in my parents' era. Trade enriches our lives every day by increasing the variety of foods available to us at affordable prices.

However, international trade is also very contentious because it generates winners and losers. In the case of the North American Free Trade Agreement (NAFTA), US corn producers were clear winners. They gained increased access to the Mexican market. On the other hand, Mexican corn producers were hit hard, with many of them having to leave the rural sector altogether and move to the already overcrowded cities. Those who remained had to compete with lower-priced exports from the United States. But in the case of manufacturing, the opposite occurred. Many assembly plants closed their US operations and moved to Mexico or elsewhere to take advantage of lower wages. This has contributed to concerns about a "race to the bottom" in which unskilled wages around the world fall to the level of the poorest countries. Of course, viewed from the perspective of those countries, this is more like a "race to the top," with wages in the coastal regions of China rising at a double-digit annual pace.

Trade is a two-way street. Opening to world markets shrinks sectors without comparative advantages and expands those that are relatively more competitive. This is a stressful process for workers as well as owners of the enterprises in the relatively uncompetitive parts of the economy, prompting bankruptcies and unemployment.

The challenge lies therein. International trade is contentious and carries some negative consequences. Nonetheless, we need it to ensure global food security. Can we move beyond the contention and successfully negotiate the international trade we will need to successfully feed the world?

A Snapshot of Global Trade Today

Given the complexity of international trade, it is no wonder there are also many ways to measure its importance. One way, appearing straightforward on the surface, is to take a look at how much of what is produced is sold abroad and how much is bought from abroad. In other words, measure exports and imports.

This is complicated by the fact that we don't always have the necessary data to perform these calculations. Farms usually don't know where their crops go once they are sold because they usually sell their grain to a "middleman." Even if they wanted to, they couldn't tell us in most cases whether their corn was exported or used domestically.

This doesn't stop us from estimating the importance of trade, it just forces us to adjust how we do so. Rather than focusing on individual farms, we turn to the national statistical offices.[1] In the United States, they report that nearly one-third of crops were exported in 2011, which is far more than any other broad sector of the American economy. What's more, this food export ratio is nearly *three times* as high as the global average for crop exports.

Global exports of livestock and processed food, however, are much lower, as a share of output, than the global average found for crops. In the case of livestock, this is due in large part to the high cost of trading live animals. It is expensive to crate a cow and ship it across the world. Furthermore, livestock products, such as fresh milk and meat, are perishable.

A country like Japan doesn't have much land for growing crops. Yet they raise beef. Cows need to eat a lot (a lot more than we do, coincidentally), and their food comes from the very crops for which Japan does not have much land. Thus Japan needs international trade in crops to feed its cows. Such trade allows countries without much farmland to import feed for their livestock at world prices.

In the case of processed food, it's a different story. Most potential trade has been supplanted by foreign direct investment.[2] US and European food companies are particularly active in this arena, with processing plants around the world. By importing raw ingredients through international trade, companies like General Mills can produce breakfast cereal and other food products at domestic facilities abroad. In this way, they avoid shipping bulky final products, while simultaneously tailoring products to local tastes and preferences, which vary greatly across the globe.

The relative importance of international trade varies widely, both across different segments of the economy as well as across countries. For instance, mining and mineral products are the most heavily traded globally, whereas services are the most lightly traded. It is not hard to imagine why—it's pretty hard to get your hair cut in India when you live in Indiana, even if the barber's fees are much lower!

The geography of international trade is, as mentioned, also quite diverse. Europe is the United States' most important export destination overall. However, when it comes to crop exports from the United States, China and East Asia are the dominant players. And East Asia is by far the dominant export destination for livestock products, followed by Mexico in second place. When it comes to other processed food, however, our neighbor to the north comes in first. Canada is the destination for more than a fifth of processed food exports.

Nonetheless, the United States has a relatively low overall share (roughly 12 percent) of exports in total gross domestic product. This is just a third of the European average and is far less than that of our North American trading partners Canada (27 percent) and Mexico (about 30 percent). This is partly due to the size and diversity of the US economy. For example, if the European Union were counted as just one country rather than 28 individual nations, their trade dependence would be counted as much lower.

The Future of International Food Trade

Although it is useful to know something about the pattern of world food trade in the present, it is more interesting to think about how this might change in the future. To answer this, we use an economic model. The model connects assumptions about population (e.g., how many people will live in that region?), economic growth (e.g., what will the average salary be?), and agricultural productivity (e.g., how efficient are farmers?) with trade outcomes. Before we dive deeper, however, let us bear in mind the wisdom of George Box, a famous statistician, who noted that "all models are wrong."[3] Yet before we give up, he did follow this pithy observation by noting that, nonetheless, "some models are useful!"

As such, let's see if we can tease out some *useful* insights from a model of global agriculture, nicknamed SIMPLE (a Simplified

International Model of agricultural Prices Land use and the Environment). The SIMPLE model was developed by my colleague Uris Baldos and me.[4,5] We use SIMPLE as a lens through which to interpret the past, such that we may be able to anticipate the future.

One of the ways to test a model is to look back at history. We can put in data and see how accurate the results are, based on what we know to have happened. We did this with the SIMPLE model over a 45-year period, from 1961 to 2006. The model "predicted" growth in global demand over that time. We know from history that global crop output tripled. However, as expected, the SIMPLE model does not get this prediction quite right. After all, it is just a simple model. However, it does come close—predicting an increase of slightly under 200 percent. But what is more interesting is how it allows us to understand the relative importance of drivers of change. The tripling of global crop output from 1961 to 2006 was unprecedented, and understanding the "whys" behind this massive growth in global consumption of crop production can give us important information about the future.

The dominant force was the growth in population. Over the 1961–2006 period, the global population grew from three billion to nearly seven billion people. The second most important driver of increased crop consumption was technology. Due to improvements in the productivity of agriculture, crop prices over this period fell, thereby permitting increased consumption. The final driver behind the tripling of global food consumption was income growth. As people became wealthier, they consumed more food—and more importantly more livestock products. Eating the animals that feed on crops is a more intensive form of consumption, one that requires more output overall to have a diet with the same number of calories.

Now here's where it turned interesting. When we looked at future projections, we first expected a rather similar outcome. After all, many economic forecasts are essentially extrapolations of past trends. However, as we began to dig deeper into the underlying drivers of global change in the food sector, we realized that the future was likely to look quite different.

The projected growth in global crop output from 2006 to 2050 shows an expected growth rate only half as fast as that over the historical period of the same length. Why is this? Population growth. It will be far less important as a driver of global crop demand as we move to the mid-twenty-first century. We already know this from my colleague's chapter on population. Population growth peaked in the late twentieth century, and the growth rate is expected to fall by half over the coming decades. Additionally, as we might recall, the population growth in the twenty-first century is expected to occur in the poorest countries of the world—many of which are in Africa—where current consumption levels are low. For this reason, adding a person in Mozambique puts less burden on global food production than adding another consumer in Indiana.

We hope that many of the poorest households in the world will be lifted out of poverty in the coming decades. This means they will be upgrading their diets and boosting overall demand for farm output owing to income growth. Therefore, by the mid-twenty-first century, we estimate that, for the first time in recorded history, income growth will become a more important driver of global food demand than population.

The implications for future trade patterns are striking. The following analysis demonstrates just how striking it will be. Figure 7.1 depicts this current pattern of international crop purchases by region of the world, which can be contrasted with the *change* in this pattern over the 2006 to 2050 projection period obtained from the SIMPLE model. As might be expected, the world's largest economies of Europe, the United States, Canada, and China dominate current purchases of crop commodities on the international market. They are expected to continue to dominate such purchases in the coming decades.

However, all three of these regions have relatively modest population growth rates, particularly Europe and China. Their economic growth rates are also slowing down, even China's. This means they are unlikely to be buying a lot more crops in the future than they do today. In fact, our projections suggest they will be buying 5 percent

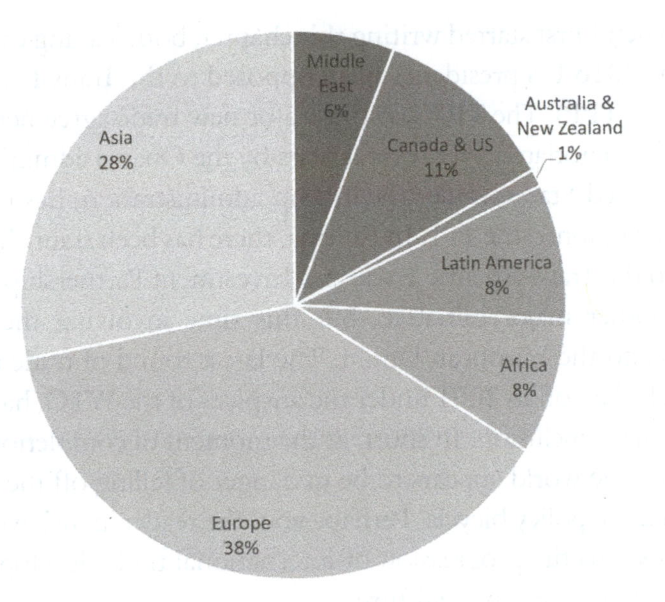

Fig. 7.1. Current share of international crop purchases.

fewer crops from the world market in 2050. On the other hand, sub-Saharan Africa has both very high population growth and accelerating income growth. This combination, together with historically low farm productivity growth in the sub-Saharan Africa region, positions that region to potentially account for about three-quarters of the net *growth* in international crop purchases. This region is followed in relative importance for future growth by South Asia and then the Middle East and North Africa (fig. 7.1).

In short, we might think of Africa in 2050 as the East Asia of today—at least when it comes to agricultural imports.

Trade Policy in Crisis

Thus far, we have said nothing about the contentious matter of trade policy. The model projections just discussed assume policy doesn't change much from the current state of affairs. However, there is a reason people say that trade policy is like a bicycle. You either pedal forward and maintain your momentum, or you fall off.

When I first started writing this chapter, both leading candidates for the 2016 US presidency were opposed to the Trans-Pacific Partnership (TPP). The TPP was the major new trade agreement painstakingly negotiated over several years by the Obama administration. The United States, under the Trump administration, has now officially abandoned the TPP. In Europe, there has been staunch opposition to the Trans-Atlantic Trade and Investment Partnership (TTIP), yet another major initiative, but this time involving the United States and the European Union. The latest round of trade negotiations, kicked off in 2001 under the auspices of the WTO, have failed to reach a conclusion. In short, at the moment of completion of this chapter, the world appears to be in danger of falling off the international trade policy bicycle. Perhaps you, the reader, may have further insights into the progression of international trade development as you set your eyes on the future.

Agriculture is at the center of many of these trade controversies. Japanese and Koreans are anxious about opening their heavily protected markets for rice. Canadians are seeking assurances that their dairy industry will not be devastated by imports. The United States continues to protect its sugar market, among others. Europeans are concerned that TTIP might allow the importation of food products with genetically modified organisms (GMOs). The European Union's approach to food regulation is diametrically opposed to that in the United States. In the United States, GMOs have been deemed safe until proven otherwise. In contrast, in Europe, new foods are deemed unsafe until *proven* safe to eat. And it is not just the wealthy countries raising challenges to free and open agricultural trade. India has been a vocal opponent of agricultural reforms under the WTO's Doha Round of trade talks, asking for "special safeguards" to prevent the erosion of farm output in sensitive (i.e., heavily protected) sectors.

These are all cases of agricultural protectionism, which has a long and colorful history. Over time, we have seen episodes of liberalization followed by periods of isolation.[6] Memories of starvation during World War II remain vivid in the cultural consciousness of many

Japanese, and this has contributed to a reluctance to rely excessively on imports. Nonetheless, over time, they have come to rely more and more on agricultural products produced overseas as their domestic farm sector has become increasingly uncompetitive.

Running in parallel to this desire—the desire to ensure adequate food during times of war and shortage—is the political imperative in many countries to protect agricultural producers. As countries develop from subsistence farming to commercial agriculture, there is a tendency to shift from taxing agriculture to providing subsidies to farmers. While this feels counterintuitive (wouldn't poor farmers need subsidies more than commercialized producers?), it is in fact the case.

At low income levels, with poorly developed fiscal systems and a majority of the population engaged in agriculture, farm commodities are one of the few places to find tax revenue. However, as countries become wealthier and farm populations shrink, the remaining producers typically become politically organized and more influential.[7] As a consequence, across a wide range of countries there is a well-documented tendency to gradually evolve from taxation to the subsidization of the farm sector. China is one of the recent and most important examples of this evolution. Over the last decade, China has dramatically increased its spending on the farm sector, while reducing or eliminating rural taxes.

Once a country has begun subsidizing its farm sector, the next step is to limit agricultural imports. How else can one protect domestic producers from overseas competition? Short of resorting to outright cash payments, that is, as they are deemed politically vulnerable and therefore undesirable (it should be noted in passing that, in Europe, the idea of cash payments for environmental amenities provided by farms is catching on).

In general, this pattern of subsidization of the farm sector prevents the natural evolution of agricultural production toward regions with a comparative advantage in farming. The only factors pushing back against this natural tendency to restrict farm imports and freeze the current pattern of production are the multilateral and bilateral trade agreements that have been negotiated over recent decades, particularly since conclusion of the Uruguay Round of the WTO.[8]

In this context, the importance of these international trade agreements cannot be overemphasized.

The Case of Dairy and the TPP

There is much to be said about international trade agreements. Many books and articles have been written about them. Indeed, I have written more than a few myself.

Not only would it be impossible to summarize them all in one short chapter, it would likely be impossibly dull to dig through as a reader. Yet a case example can give us considerable insight into the matter of trade and agriculture. As such, let's dig into the particulars of one such case, a relevant one—that of dairy products under one of the current trade agreements under negotiation, the TPP.

Two of my former students, now gone on to spread their academic wings, made a case study of this particular topic. These two, Jason Grant and Everett Peterson, both now at Virginia Tech, in addition to Sharon Sydow, a member of the chief economist's staff at the US Department of Agriculture, analyzed the dairy sector and international trade in depth, a task more complicated than it may superficially appear.[9] When it comes to international trade, the dairy sector is surprisingly heterogeneous.

There are two dozen types of dairy products traded. These often have dramatically different tariff structures. This can depend on the competitiveness of the sector, as well as its political influence on trade policy. In the United States, for example, imports of fresh cheese face a relatively high effective tariff. Milk powder, skim milk, and casein (a protein by-product of dairy processing) are imported with virtually no barriers at all. In addition, dairy imports are riddled with something called the tariff rate quota (TRQ) system.[10]

TRQs are essentially a means of regulating trade while avoiding the use of import quotas. Import quotas were banned under the Uruguay Round Agreement of the WTO. Under the TRQ system, a prespecified level of imports is allowed to enter the market at a relatively low tariff. However, once a certain level of "in-quota" imports is reached, the tariff jumps to a higher, often prohibitive, level.

TRQs result in a lot of side deals that are often bilateral, and these arrangements may also be overly influenced by nontrade geopolitical factors and relationships. In a rapidly evolving global economy, this type of prespecified sourcing of products is bound to result in some pretty dramatic inefficiencies. Of course, if these inefficiencies were eliminated, as is the goal of many trade agreements, it would cause a major restructuring of the industry. One such example is offered by New Zealand's dairy product exports to Canada and the United States, where regulation has resulted in high dairy prices and an industry that, on average, is not internationally competitive. Opening the border to dairy imports from New Zealand, a highly competitive producer, would result in many small dairy farms in Canada and the United States going out of business. It would also result in lower revenues for all domestic dairy producers. By maintaining a TRQ on dairy imports, with a high out-of-quota tariff (51 percent on butter imports from New Zealand into the United States), the government is able to protect its domestic dairy industry.

Grant, Peterson, and Sydow emphasized the US export view in their analysis. They found that the United States would gain $63 million in welfare by implementing the dairy component of the TPP. This stems largely from increased profits to the dairy sector due to improved access to the Canadian and Japanese markets (also Vietnam, in the case of butter and skim milk powder). These producer gains are somewhat diluted by the higher prices that consumers face in the United States as overseas demand picks up as a result of the TPP. The authors contrast these gains with the outcome in which the United States does not participate in the TPP. In this case, the dairy industry is estimated to lose $11.5 million.

Of course, dairy is just one of hundreds of sectors in the US economy. A comprehensive trade agreement like the TPP would affect nearly all of them in some way. Not all sectors will be winners, as the political debate has shown. Many of the labor-intensive light manufacturing sectors will likely experience further contraction and job losses. Production will move to parts of the world where labor is cheaper. The TPP went well beyond just tariffs, to encompass issues ranging from intellectual property (a topic of keen interest to the pharmaceutical industry), to rules governing foreign investment, labor and environmental standards, and so on. It is no wonder that, with so many diverse interests across sectors and countries, negotiating such an agreement takes years.

With the United States opting out of the TPP, this is a significant setback to the forward momentum of the "international trade bicycle." Given the strong momentum toward economic integration in East Asia, this agreement may proceed nonetheless, even without the United States. It is also likely that, in the absence of US leadership, China, as the world's second-largest economy, will move into this vacuum. China is not currently part of the TPP. Indeed, much of the agreement has been designed to goad China into reforming its own domestic policies. If the TPP fails, regional trade policy in Asia will likely move in a rather different direction, with far less emphasis on opening agricultural markets and more emphasis on development of Asian manufacturing supply chains.

Life in a Globalized Economy

In the current political environment, international trade agreements play the scapegoat. They are blamed for a whole host of ailments. However, the irrefutable fact is that the global economy is becoming far more tightly integrated. In the words of popular author and columnist Thomas Friedman, "the world economy is flat!"[11]

In this kind of highly competitive environment, any slight advantage one country may have can allow them to dominate the global market. We've seen this in the case of cell phones, solar panels, and other specialized manufactured goods. Agricultural production is somewhat different. It requires suitable land and climate, factors that cannot themselves be "produced." And these natural resources are not evenly distributed around the globe.

For this reason, the topography of the global agricultural economy is not nearly as "flat" as it is for manufactured goods. Nonetheless, countries have a lot of influence over the success of their agricultural sector. Investing in knowledge and infrastructure can boost future productivity in big ways. This has historically been the case for the United States. The future, however, depends on us. If the current slowdown in agricultural research and development in the United States continues, we will likely grow far less rapidly than other countries, most notably Brazil, China, and India, all of which have accelerated their investments in research and development.[12]

However, agricultural trade is about far more than just getting a leg up on the competitor. It provides something else that is extraordinarily important—flexibility. Trade gives us an important buffer against unforeseen weather events. In the context of ongoing climate change, international trade will become increasingly valuable in the future.

Take the heat wave and drought of 2012. It prompted major losses in corn harvests in the United States. We lost more than a quarter of our corn crop that year. As a result, in the United States itself, corn prices rose sharply and domestic use of corn was reduced. Exports also fell. Consumers around the world, as well as producers in the Southern Hemisphere, had to adjust to the shortfall. These adjustments served to moderate price increases in the United States, but this was only possible because of international trade. Not every region on Earth that grows corn experienced a similar drought that year, allowing others to step in to fill the gap. Similarly, during the 2001 and 2002 drought in Australia, irrigated rice production in the Murray–Darling basin was temporarily eliminated. Australians didn't starve—they imported rice from elsewhere.

In the face of climate extremes, trade is as an extraordinarily important buffer. For international trade to play this role, however, it must be free to do so. This depends critically upon the "rules of the road." Governments tend to try to adjust their agricultural trade policies in order to insulate their domestic markets from international price changes, particularly during times of extreme price spikes.[13]

This has a range of undesirable consequences, beyond blocking the flexibility we need to buffer our food supply across the globe; such policy interventions can further exacerbate world price volatility. For example, during the 2006 to 2008 price spike for rice, this kind of insulating behavior accounted for *nearly half* of the overall increase in the world price of rice.[14] The figure for wheat was nearly one-third. The TRQs mentioned earlier, in the context of dairy, make the problem even worse. When a country's imports are facing a binding TRQ, there is no incentive to adjust domestic consumption or production in the face of a global shortfall. This forces ever greater adjustment onto the countries, often the poorest ones, without such restrictive agricultural trade policies in place.

Without ways to stop counterproductive trade policies, the world is in for a rough ride in an era of increasing crop supply shocks due to climate volatility. Yet what about the downsides to international trade? What about the losers it creates?

The answer is, yet again, flexibility. This time, however, it isn't the flexibility to import and export food products as needed. It is flexibility in our labor markets. Workers must be nimble and adaptable, able to move from one role to another as sectors grow and shrink, and as jobs are won and lost.

Evidence suggests, however, that this is easier said than done. In an extremely influential study of the impact of China's emergence as a global trading superpower on the US labor market, researchers found that workers in many parts of the United States were remarkably slow to adjust to the new economic landscape.[15] This resulted in depressed local wages and high unemployment in regions with industry exposed to competition from Chinese imported goods.

These findings make it clear how "short-run gains" from freer trade are exaggerated in most economic analyses. They typically assume that there is a great degree of labor market flexibility. In this context, it is not difficult to understand the disconnect between economic analyses of the gains from trade and the groundswell of opposition to current trade agreements. It is no surprise, either. How would any of us react if we were promised rewards for years of loyal service to our employer, and then found ourselves out of a job with nowhere to go?

There is a way to manage this. We need to invest in creating greater flexibility for our workers. One reason displaced workers have a hard time finding new employment is limited training. A high school education is not enough to ensure meaningful employment in the twenty-first century. Improving our educational system, not only for youth but also for adults, is a necessary step forward for greater labor market flexibility.

Divorcing health care and retirement benefits from employment would also help make workers more nimble. Many people hesitate to change jobs or adapt their profession because of a well-founded fear of losing such critical benefits. Creating a safeguard so benefits move with workers when they change jobs is not only a valuable safety net but also increases the speed with which workers can adjust.

Changes are coming, and more rapidly than before. Our economy is increasingly globalized, and fighting globalization will hurt ourselves, and our food security, in the end. If we invest in our ability to adapt, this will, in turn, allow for the full economic benefits of international trade to be realized. We can take the benefits of globalization and funnel them into investments in our most prized commodity—*people*.

Chapter 8

Spoiled, Rotten, and Left Behind

Ken Foster

Not everything we grow is eaten.

Back in the 1960s when I was a kid in rural Indiana, we lined up at the end of lunch period to dump our leftovers into a big plywood box with three holes cut in the top. One by one we filed past, scraping the uneaten items into this contraption before running off to recess. It was just what we did. Who knew or cared where all that gross slop went, anyway?

I did, actually. I cared far more than I wanted to. My mother was a teacher and my father was a farmer. It was the perfect vantage point from which to know *exactly* what happened under those holes. There were three big, shining metal garbage cans. And every day after school my brothers and I loaded them, filled with something that looked, to my eight-year-old self, a lot like barf, into the back of the family station wagon. We then rode home to our farm where these leftovers were fed to a small herd of pigs. And they could not have been happier pigs, by the way, feasting on a well-balanced diet designed by the school nutritionist for growing kids.

Reflecting on this recently, I stumbled upon one of my former schoolmates. I decided to ask him where he thought all that went. "Hey, do you remember that big plywood box where we dumped our food after lunch? Where do you think that stuff went?"

He got a puzzled look on his face, perhaps due to the odd nature of the question itself, or perhaps his inability to answer, and said, "I guess some sort of garbage disposal system?"

If he had thought about it, which we normally don't, he would have known this wasn't correct. The box didn't sit there all day, and when it wasn't there, there were no pipes or holes in the floor or a disposal system. So I enlightened him as to where it really went, and he laughed.

Today, feeding food refuse from an elementary school directly to livestock would be a food safety violation. People would frown on large cans of slop sitting around with the potential to draw flies and other pests to the school. As a result, schools have installed modern garbage disposal systems, such as the one my friend had envisioned. Wasted food is sent to the waste treatment facility or the landfill, as is the case in many of our homes.

My old schoolmate's response highlights a poignant point; we, in the developed world, don't think about this stuff (unless, as in my case, you had to). Yet, despite our ignorance, willful or otherwise, the fact remains that the great quantities of food that go uneaten are a huge, huge problem. As the title of this chapter hints, it is far from a glamorous subject. It is in fact a distressing one, and one we must confront if we wish to safely and sustainably feed our *whole* world.

This is exactly what we look at here in this chapter; all the food that goes uneaten, and why. Simply put, food goes uneaten for two reasons. It is either wasted, or it is lost. It is easy to confuse the two terms—"food waste" and "food loss"—as being synonymous. But it is essential to understand them as the two completely separate phenomena that they are, and to grasp the differences in their causes and potential solutions.

The difference between "loss" and "waste" depends primarily on where it happens in the supply chain. The supply chain is the very long and complex process that runs all the way from when a farmer plants a seed or breeds livestock to when a person takes a bite of that food. Think of it as a literal chain, with various links all hooked together, that "supplies" us with our food. This supply chain can include dozens of handlers, processors, and marketers, depending on the nature of the beginning raw material and the final products. It represents the combination of important resources like water, land, air, and human labor. Chapter 6, on systems, explores the many ways these supply chains manifest in our world, such as organic farms, conventional farms, farmer's markets, and so on.

To better understand the nitty gritty details of a supply chain and how it relates to food loss and waste, we can look at the example of corn. The very first link in the supply chain is out in the field where the corn is grown. The farmer plants the seeds, carefully tends to them, and at the end harvests the corn itself. Along the way, however, a part of the crop is lost. For instance, the wind comes along and flattens the corn stalks. Or sometimes some of the corn gets dropped as the harvester runs, particularly when the machine takes turns at the end of each row.

This is what we call harvest losses. This is a part of *food loss*, the very first place where food goes uneaten along the supply chain.

The next stop along the supply chain is the storage and transportation process, when corn is usually stored, sorted, transported, and processed. Each corn-based product follows its unique supply chain toward, for instance, a retail establishment like a restaurant, grocery store, feed store, or gas station. Along this part of the process, more food is lost. This is postharvest loss. It happens after the food is harvested, but before it makes it to anyone's dinner table (or feed bin, or fuel tank).

Together harvest losses and postharvest losses make up the entirety of food loss. It is all the bits of food or potential food that are grown but simply don't make it to the point where they can be used for their intended purposes. All of this occurs in the beginning and middle of the supply chain.

Food waste, however, happens at the very end of the supply chain. This is any food that is discarded, for any reason, at places such as shops or restaurants or school cafeterias, or, pretty commonly, in our homes. The slop I rather begrudgingly took home every day to feed the pigs would have ended up as food waste had it not been for my father.

Economists at the US Department of Agriculture's (USDA's) Economic Research Service have estimated that, in 2010, food waste in the United States totaled 29 percent of value (what we paid or would have paid for it when we bought it) and 33 percent of the calories that were delivered to retailers.[1] That is roughly one-third of all money spent on food and all the calories of said food. And it all just went to waste. That's a staggering, and alarming, amount of waste in terms of cash and uneaten food.

The Environmental Protection Agency (EPA) estimates that food waste in the United States is the largest single component going to municipal landfills[2] and, according to the Food and Agriculture Organization of the United Nations (FAO), every year consumers in developed countries waste more food than is produced in all of sub-Saharan Africa.[3] Natalie Donovan, my awesome graduate research assistant, sent me some calculations on the USDA numbers mentioned above. We painstakingly sorted through them together.

We reckon that, in 2010, the total food waste in the United States resulted in 364 billion lost calories per day. This is a very large number. It is hard to envision exactly what this means. Imagine if a person is eating an average of 2,000 calories a day. The total wasted calories in the United States (in one single day) could feed that one person for about 182 million days. That is a total of almost 500,000 years, which is a lot longer than any of us are going to live. In fact, it's about 5,000 times longer than you're going to live.

So the wasted calories, on a single day in the United States in 2010, could have fed you for your entire life . . . times 5,000. If you accumulate these wasted calories over an entire year, it amounts to enough calories for more than 70 billion people and has a purchase value of $162 billion.

Not all of this food can be salvaged. Not all of this food *should* be salvaged. Some of it poses food safety risks, or tremendous logistical hurdles, or it may not be nutritionally balanced for human beings. But the magnitude of just this waste, just in our own country, is staggering and grows even larger when extrapolated to other developed places in the world that have food waste profiles similar to ours, such as Canada, western Europe, Australia, and New Zealand.

If we had plenty of food in the world, then perhaps we could afford to waste this much. Yet this isn't the case. According to the USDA's Economic Research Service, in 2015 almost 16 million households (12.7 percent of total households) in the United States were food insecure.[4] To be food insecure means that a household faced difficulty at some time during the year in providing enough food for all household members. Among these food-insecure households were 6.3 million households deemed to have very low food security. This manifests itself in disrupted normal eating patterns, reduced food intake, and sometimes the inability to purchase food at all.

Worldwide, the FAO estimates that 795 million persons were chronically undernourished between 2012 and 2014. This means globally, more than twice the number of people living in the United States don't have enough to eat.

We don't have enough food to feed the world, yet based on how much we're wasting, it certainly seems we ought to. Geography plays a big role in this. Where you live will largely influence if you are wasting food thoughtlessly, or are keenly attuned to losing food despite your best efforts to preserve it. Developed, wealthy countries, like the United States, Canada, Australia, among others, experience a higher prevalence of food waste. Developing, poorer countries, such as India, Mali, and numerous others, suffer more from food loss.

Figure 8.1 illustrates this for cereal crops. The dark gray indicates data from South and Southeast Asia. This is an area with, on average, higher levels of poverty. The light gray represents North America and Oceania, wealthier countries with, on average, lower levels of poverty.

When you really stop to think, it's not too hard to understand

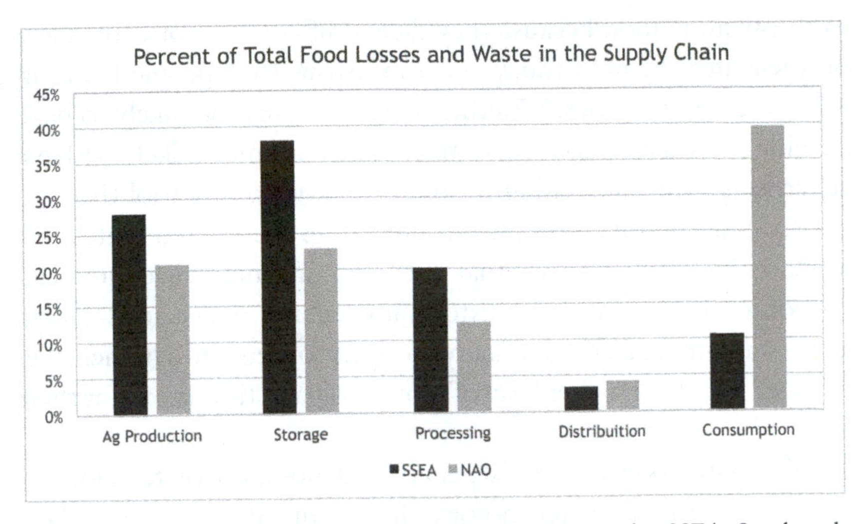

Figure 8.1. Food waste and food loss trends across geography. SSEA, South and Southeast Asia; NAO, North America and Oceania.

this dramatic contrast in the distribution of food waste and food loss between developing and developed countries. The nuances of the issue are complex. Simply put, however, there are two particular issues that come into play: a general attitude toward food (related to waste) and access to technology (related to loss).

A family on the brink of starvation will not waste a single kernel of corn if they have the means to prevent it. A family with plenty, who has never experienced real need, will more easily toss the expired strawberries or leftovers. For food-secure individuals in a wealthy country, like the United States, we are fairly certain that there will be food available for purchase in the near future. So we put more than we can eat on our plate or in our grocery cart, then we toss out our wilted lettuce or loaf of bread with the expired "use by" label. People living in a developing country, with a lower likelihood of finding food in the near future and probably no use by date, will use most of their food. They are more likely to hang onto their wilted lettuce and stale bread, and far more likely to be eating them in the near future. In fact, that lettuce probably never got to the wilting state, nor the bread to the stale stage, in the first place.

In addition, people in higher-income countries are more likely to throw away food because they face a different set of constraints on their lifestyle, both today and tomorrow, than people living in lower income countries. Wealthier countries are more likely to have labeling laws than poorer countries. They are more likely to have advertising campaigns using freshness as a marketing tool than are poorer countries. As a result, our culture in the United States, for instance, has evolved to one that does not fully internalize the social cost of food waste. To some extent, it even encourages it. Retailers and restaurants realize that quality competition leads to a higher level of waste but also a higher level of reputation and thus profit for their business.

Yet, with food loss, it is generally a question of technology. Countries with more technology, higher-quality harvesting, better transportation systems, and more effective storage systems have less food loss during harvest and postharvest periods. A large-scale farmer in Iowa will have a significantly lower percentage of loss than a poor, small-scale farmer in Ethiopia. The Iowa farmer has access to mechanized harvesting tools, effective fertilizers and pesticides, as well advanced systems of irrigation and grain storage. The Ethiopian farmer may have none of these.

The end result is that those who need the food the most will lose it. And those who have it, will waste it.

Waste

I assume most of my readers do not have a small herd of pigs like we did, where they can recycle all their wasted food. I also assume that most readers aren't purposely wasting food while others go hungry, but the brutal truth is that to a starving individual we might as well be doing so. The reality, however, is that those of us in developed countries who are used to having plenty don't generally think about it. Yet we *must* think about it. We must think about it in the long term: strategies, investment, and innovation. We must think about it in the short term: immediate behavioral changes, awareness, and intention.

The first step is, quite simply, becoming aware of the massive amount of food that goes to waste. The immediate second is setting an intention to do something. The subsequent others can be as infinite as your ability to innovate (if you're an optimist like my co-editor). Yet, even if you're a pessimistic economist like myself, there still remain many ways to prevent food waste. And, although we will never eradicate all food waste (nor should we try—the cost of doing so would outweigh the benefit), we are nowhere near the *optimal amount* of food waste. We are far, far above it.

If you travel to the USDA's Food Safety and Inspection Service (FSIS) website (fsis.usda.gov) you will eventually discover that the use by and sell by dates that we see on many food products are neither required (except for infant formula) nor are they a safety date. To coin a phrase, "They are more like 'guidelines.'"

As I write this, I imagine there are a lot of confused readers who thought that food past those dates was deemed unsafe or otherwise unfit for consumption, and consequently believed that throwing it away was the only responsible thing to do. A sell by date is set by food manufacturers as a guide to retailers. It is an estimate of how long a product should be displayed for sale. In general, you should buy the product before the date expires. But even if it is, it may still be usable if it has been processed in some way. It also likely doesn't belong in the landfill.

The "best used by" date is the recommended date for getting the best flavor or quality. The use by date is the recommended last day for use while at peak quality. Neither are food safety guidelines, nor in many cases even close to the last date that the product can be safely consumed if properly stored or processed. Yet, in many US households, food is routinely discarded after the use by date has expired. Not because it has gone bad, but because we think that is what these unclear labels are telling us to do.

This is not a recommendation to ignore these dates altogether. You should try to purchase products before any of these dates expire. Before you purchase it, the food waste is controlled by the retailer. Yet once the food is in your grocery bags, the waste is entirely up to you. When my brothers and I had food left on our plates, my grandmother used to say "your eyes were bigger than your stomach." It was her way of admonishing us for taking more than we could eat. Perishable products like meat, dairy, fruits, and vegetables should be taken home immediately and refrigerated to maintain freshness and inhibit microbial growth. In the next few days after the purchase of a perishable food, as that use by date gets closer and closer, consumers should begin to freeze anything they do not think will be used by the date. Frozen foods can be kept safely virtually forever. At some point, the quality will begin to decline, but it will still be safe to eat. These things can be added to stews, soups, or stir fries. They can add texture and color even if their flavors and nutritional values have declined. They will absorb the flavors of spices and other items that are added fresh to the mix. You may even discover that doing this can offset a flavor that is too strong.

In chapter 6 on systems, my colleagues mentioned the FLAVR SAVR® tomato developed through genetic modification techniques by the company Calgene. The modification is directly related to food quality and waste. This particular variety of tomato has a gene inserted that stops production of a specific enzyme. The enzyme it blocks is the one that causes the tomato to soften and rot. What does this mean for food waste? These tomatoes can now be picked ripe, transported in ripened form, and refrigerated for longer periods without spoiling. They are thus less likely to go to waste. Additionally, other tomatoes are often picked green, then treated with ethylene gas to cause ripening. This is why the tomatoes you buy at the local farmer's market tend to taste a bit better (by most people's assessment) than does the FLAVR SAVR tomato. In a sense, its true use by date is extended.

A few of the more commonly purchased and wasted food items are fresh meats. The USDA FSIS provides guidelines on their website for how long products can safely be stored in your home, after purchase, before you should eat them, freeze them, or otherwise preserve them. According to the FSIS, properly refrigerated fresh poultry is good for one to two days after purchase, whereas fresh red meat properly refrigerated is good for three to five days after purchase. Things like organ meats and uncooked hams and sausages are more like poultry, because their ability to sustain rapid microbial growth is greater than that of red meat.

I think we have a tendency to treat eggs like meat, probably because they come from chickens. But eggs are an amazingly stable food. Properly refrigerated eggs can last from three to five weeks. Processed dairy products are remarkably stable in the refrigerator too. Foods like cheese, yogurt, and cottage cheese have already been subjected to a controlled microbial process, and everyone knows when the milk has gone bad. When my children were growing up it was never around long enough to spoil, but now that they are grown, I have learned to detect the smell of sour milk. And that spoiled milk in your fridge is a great replacement in baking for buttermilk, yogurt, or similar ingredients. You can freeze your sour milk if you don't have an immediate need for it, and then thaw it on baking day.

What's more, people just don't like oddities in their food. We want the perfectly conical strawberry and the cucumber without the shriveled curled tip (that is caused by inadequate pollination, by the way). A few years ago, I grew several hundred pounds of potatoes of patriotic red, white, and blue colors, only to find that selling them wasn't so easy. The local stores had contracted with other farmers for their potato needs. As such, I turned to my oldest brother who grows and sells produce at farmer's markets for a living. When he saw my potatoes of all different sizes and shapes, he too declined to purchase them because his customers at the markets want uniformity of size and shape—not a hodgepodge of small and large, smooth and knobby. In commercial-scale agriculture, consumers' demand for uniformity isn't a big problem because the so-called ugly fruit and vegetables can be sorted into the processing sector to be rendered into juice, jam, and the like. But at the local scale, agglomerating sufficient quantity for profitable processing is a challenge. Maybe we can all try to be a little less picky in our local food settings and embrace the variety that nature creates—although my friends and family who received gifts of potatoes a few years ago might feel otherwise.

Speaking of food processing, I grew up canning and freezing the meat, vegetables, and fruit we grew on the farm. But most people didn't have that experience. Canning is an option for reducing food waste, but it takes some specialized equipment and sufficient quantities to fill the jars. Canned products don't keep as long as frozen ones, and many people find canned food overcooked when it comes out of the jar. Trust me, after many hours in a hot kitchen doing triage on vegetables, I can tell you that the ubiquitous refrigerator freezer is the way to go for small household food rescue missions. As my colleague points out in the next chapter on nutrition and healthy eating, you probably can't do everything all the time, so approaches that minimize effort and time spent for food rescue are really key.

Most things can be adequately frozen in a burpable plastic food storage container, in double plastic bags to reduce freezer burn, or in glass canning jars. Most vegetables and some fruits need a little extra preparation before freezing to retain as much quality as possible. You may even want to cook meat before freezing it to make it more convenient when you are ready to use it. There are many great resources on the Internet to help you learn how to preserve food safely and maintain its quality.

This is all to say that there really is no reason for the amount of food waste in American households given our incomes and access to freezing. It doesn't have to stop there. Consumers have a vibrant, strong voice in food production today. We can ask the managers of our food retailers if they have a strategy for reducing food waste. The strategy of freezing isn't typically an option for retailers, but most towns and cities have food banks and homeless shelters that do an amazing job of utilizing items that are nearing their expirations. The retailer should be able to obtain a tax benefit from these in-kind charitable contributions. If you find that your retailer doesn't have a strategy for minimizing food waste, then you might suggest a local charity to them and encourage a partnership.

As I alluded to earlier, sometimes there is just going to be some food that can't be used. People like to peel their potatoes and cucumbers, and many fruit peelings are not edible. There are times that fruit spoils before we can get it into the freezer. We sometimes go away on vacation only to, frustratingly, return to a broken refrigerator. What do we do about these unavoidable food wastes?

I still live in the country, so our fruit and vegetable peelings and the occasional spoiled items are mixed into the soil in our garden. And although not everyone has a garden, nearly everyone has access to some patch of soil or grass that could benefit from some organic matter. The beauty of composting is that it is a scalable technology. The biggest composting bin is limited only by your ability to occasionally mix the contents, and it scales down such that you could compost in a thimble if you wished. All you need is a sealed container and some soil. Put some soil in the container and whenever you have a banana peel or similar item, just toss it in and shake the container. I like a bucket because you can just turn it on its side and roll it around to put the food waste in contact with the soil. The microbes in the soil will do the rest. If you see the food waste to soil ratio is getting high, you can add more soil. After a month, you probably need a second container but keep mixing the first one. Eventually, the food waste will become soil, and you can pot some flowers in it or spread it on the lawn. An insider tip, keep your composter in the garage or on the deck so that when you open it to add food, no one has to be exposed to the potential aroma.

As I said, composting is a scalable technology. In theory, there could be a community composter for a whole town. How is all of this food waste happening, all over the cities and communities, going to be collected in one place? Here emerges my inherent pessimism. But the next story is an example my coeditor loves, because it highlights the potential of what we could do if we tried.

In September 2015, the governor of New York, Andrew Cuomo, announced that the Long Island town of Yaphank would become the location of the world's most sophisticated food digester. The plant aims to convert 160,000 tons of food waste and grass clippings into biofuel and organic fertilizer. If all goes as planned, the resulting biofuel will power the plant, as well as the trucks that deliver fertilizer and collect food waste.

This is a remarkable idea, despite not being entirely new. In fact, it isn't that different from my father's pigs and the food waste from the local school. His pigs, not unlike the Yaphank digester, turned food waste into organic fertilizer and meat. In 2010, at Purdue University where I work, they began a collaborative project with the city of West Lafayette, Indiana. Food waste from the campus food service facilities is collected and transferred to the city's waste treatment plant where it is digested into organic material and biogas. The biogas supplements the energy needs of the digester and the city wastewater treatment facility. While less efficient than the proposed plant at Yaphank, it has reduced natural gas purchases of the waste treatment facility by an estimated 60 percent and is making use of the food wasted by a large group of students residing in Purdue's dormitories.

Loss

Imagine that you and your family have plowed a small plot of land using hand tools and hoes. Together you planted a staple crop like corn or cowpeas and tended it caringly by hand-pulling weeds and carrying animal manure for fertilizer. Finally, at the end of the season, you are rewarded with an excellent harvest. The crop is destined to feed your family. What is left over will be sold to pay fees for your children to go to school.

Now that your family has its bountiful harvest, what should they do with it? It is likely that everyone else in your community experienced good yields too. If that's the case, the price of grain is extremely low at harvest time, right when you would want to sell it. When this happens in the United States, farmers store their crops in large ventilated structures sealed off from pests and optimized to prevent the growth of molds so they can sell later when the prices are higher.

Yet on smallholder farms in a developing country, farmers don't have access to those sorts of storage technologies. They can sell their crop for a low price at harvest, or they can store the crop. But if they store the crop, they will suffer huge losses due to insects and mold. Just in sub-Saharan Africa alone, the value of postharvest grain loss has been estimated to be as high as $4 billion. All over the developing world, losses to insects range from 30 to 50 percent. The crop that remains after storage is low in quality because of the insects and their droppings. The damage done to the grain lowers nutritional value and can lead to poor flavor. The alternative is insecticide, which is often ineffective and has the potential to poison people.

So in the end, storing the grain isn't really an option for these farmers. They are left with the choice to sell the proceeds right at harvest time when the price is the lowest.

This is the type of farming situation most vulnerable to food loss, and it is indeed most common in developing countries. The challenge for these farmers is huge. It affects all aspects of their lives from the food security of their family to their children's education and things we would never have imagined. The situation above described a scenario in which a family was lucky and the weather was favorable. Yet extreme weather events like long-term drought and hurricanes can wreak havoc on the crops of smallholder farmers. Even in good years they do not have modern mechanical harvesters that minimize grain dropping on the ground and facilitate quick harvests before birds and other pests can reduce yield. And as we saw, trying to store the crops can lead to huge losses.

My first experience with the importance of storage in developing countries was as a Peace Corps volunteer in Central America. Many of the rural villages in the early 1980s lacked any modern conveniences. A group of British military personnel started a project to build latrines in some of these villages because of the potential for a disease cycle between people and their semiferal pigs. When we returned a year later, we found that in many cases the latrines were full of corn because they provided protection from rodents and rain. Immediate protection of the grain was more important than other long-term health concerns.

My Purdue colleague, entomology professor Dr. Dieudonné Baributsa, grew up in the Democratic Republic of the Congo. Food insecurity and hunger are no strangers to him. Today, he focuses his research on a particularly hideous form of food loss in developing countries—when insects eat the food that people need to survive.

People in developing countries are literally dying for the technology to protect the harvest in which they have invested so much of themselves. In technical terms, we know this is called postharvest loss. Dieudonné has focused his entire career in this area, particularly on development and education about hermetic storage. The origins of the word "hermetic" are less scientific than the word sounds. It traces to the Greek god Hermes or Hermeticus. Hermes was the god of alchemy and astronomy and the word "hermetic" (meaning airtight to us today) arose because Hermes was purported to be able to seal a box in such a way that it could never be opened.

Can you make the connection? Dieudonné is developing ways to store foods and educating farmers on how to store their grain so it isn't eaten by insects. It's a system that prevents bugs from getting in, and one in which the bugs that are already present will asphyxiate from lack of oxygen. There are a number of such technologies because any container that can be sealed airtight will work. However, some may work better than others in the developing world, where the amounts that farmers store are variable from year to year, and there is very limited cash to invest in storage containers.

Imagine a metal or plastic drum with a lid that can be sealed. If the farmer needs to seal just the amount that the drum will hold, then it will be great. But if the farmer has less than enough to fill a drum, then there is no way to get the air out and suffocate the insects.

Emeritus Purdue entomology professor Larry Murdock solved this problem when he designed the Purdue Improved Crop Storage (PICS) bag. It is a triple-layer, heavy plastic bag that, when properly sealed, will result in maintaining very high grain quality. The beauty of the PICS bag is that, like the home composter, it is scalable. One can make the container as big as the equipment available to handle it or as small as is needed. The usual PICS bag holds 100 kilograms of cowpeas, but if a farmer has half that amount the bag still works because it can be squeezed down to whatever amount is available, and the air can be forced out to suffocate any insects. As of June 2016, over seven million PICS bags had been sold in Africa and Asia. Each bag is estimated to add $150 to a farmer's income—more than a 30 percent increase for many of them.

This example drives home the point that technology is the needed element in solving food loss. It may be technology in the United States that reduces losses during processing or harvesting activities. It may be hermetic storage that protects grain quality and allows farmers to market their grain at higher prices later in the year. Either way, the answers arise from the ingenuity of people like Larry and Dieudonné and others.

The Challenge

Solving the problem of our uneaten food, both from loss and from waste, is a critical component of feeding our world. As we have seen in earlier chapters, the stakes are extremely high. Even today, many people don't have enough to eat. As the global population grows and climate change steadily increases its impact on our food production, the pressures will expand. It is also not just people who are directly harmed through uneaten food; the environment is harmed as well. All that wasted and lost food represents tremendous efforts by farmers and a major depletion in natural resources, such as soil and water.

The situation today is far from ideal. Yet there is tremendous potential for reduction in food loss through education, technological innovations, and development. There is tremendous potential for reduction of food waste through awareness and concerted efforts to redirect waste toward innovative programs. We have the wherewithal and the technology, as the people in Yaphank, New York, and sub-Saharan Africa are discovering. We just need the collective will to implement and educate.

Chapter 9

Tipping the Scales on Health

Steven Y. Wu

Modern society's untidy balance of costs and benefits for staying healthy

Eating is not easy when dinner is supposed to satisfy hunger, prevent metabolic syndrome, and save your community *and* the world. So maybe you decide to focus. You narrow your list of requirements for dinner down to just health. Ignore the other stuff. This should make things simpler, right?

So then you go online and soon discover that all the various and sundry food "experts" or "gurus" have wildly differing opinions about what is healthy. Some vegans tell you that cancer and heart disease await if you so much as touch any animal products or fat. The low-carb gurus explain how eating carbs spikes insulin, which in turn causes you to store fat and be at risk for metabolic syndrome. Yet another group of foodies assures you that eating local and organic is the key to health.

Thoroughly confused, you check the science. But wait, there's a myriad of new findings about the harmful effects of various macro- and micronutrients. Then some scientists say carbs are bad for you, while others say they're good. Sugar? This week it's okay. The next a nutritionist defends its use but *only* in moderation. Another study says sugar (glucose) fuels the growth of cancer cells. But then a recent article shows that it is actually protein (amino acids) that fuels cancer cell growth. Later you discover a new finding proclaiming you need more salt, even though avoiding too much salt has been the standard guideline for decades.

Right. So basically, there seems to be scientific evidence to show you that anything and everything can kill you. You try to follow the government guidelines to eating healthy and going on a diet to lose weight, but it's not easy. Not easy at all. Too many rules to follow. Too difficult to hold back those hunger pains by cutting calories. You read that the overwhelming majority of people who diet tend to regain the weight they've lost over the long run. Overwhelmed, confused, frustrated, and seriously annoyed, you are on the verge of giving up, and, like many people, maybe you do.

I would hazard to say that there are precious few people who, living in countries that afford easy access to food or having the socioeconomic status to access food, haven't furrowed their brow in confusion and frustration over what to eat. Add to that a growing food movement that seeks to make food inseparable from health, morality, environment, and community, and the situation has become unmanageable.

The problem isn't just our peace of mind. On the contrary, confusion over how and what we should eat causes a chain of damaging effects. Our poor diets have led to skyrocketing levels of obesity and related health problems. Not only does this severely decrease our quality of life and put us at risk of early death (as if this isn't enough) but it puts a massive financial strain on health care systems and diminishes our ability to be productive, working members of society.

"Apart from tobacco, there is perhaps no greater harm to the collective health in the U.S. than obesity. Worldwide, too, obesity's health effects are deep and vast—and they have a real and lasting impact on communities, on nations, and most importantly, on individuals, today and across future generations," explain the authors on Harvard's School of Public Health website. Obesity, they go on to explain, causes or is closely linked with a large number of health conditions, including heart disease, stroke, diabetes, high blood pressure, unhealthy cholesterol, asthma, sleep apnea, gallstones, kidney stones, infertility, and as many as 11 types of cancers, including leukemia, breast, and colon cancer. No less real are the social and emotional effects of obesity, including discrimination, lower wages, lower quality of life, and a likely susceptibility to depression.[1]

Excepting certain cultures where obesity is considered a sign of wealth, no one really *wants* to be overweight. Yet, between 1960 and 2008, obesity in US adults increased from 13 to 34 percent. This shift is so severe it has been called the obesity epidemic. Over a third of the US population is obese, and this is not isolated solely to this nation. According to the World Health Organization, worldwide obesity has doubled since 1980. In 2014, 39 percent of adults over 18 were overweight. And, incredibly, *most of the world's population lives in countries where overweight and obesity kill more people than underweight.*[2]

In feeding the world, it is not enough to simply have the calories available. It is the exact form those calories take and the costs and benefits associated with choosing them. This can be looked at from two perspectives. The first is from the viewpoint of those of us living in socioeconomic conditions that allow us access to food. In this instance, what we are driven to eat, and how these choices influence our health, is of tremendous importance. It isn't just the fully stocked shelves that matter, but what those shelves contain and, even more, our ability to determine just what to choose among the options they display.

The second perspective is from those people who don't have access to the right foods at all, even if they want them. My colleague addresses this in the last chapter in this book, where he explains how many people can't get good nutrition because of their circumstances. His chapter is about achieving equal access around the globe, which is a different and crucial issue. For now, though, we'll focus on the challenges of good health where there *is* access to food, and the challenge of navigating these choices so we can tip the scales of health in our favor.

The Economics Approach

You have probably tuned in to Bloomberg or CNBC and heard an economist with an impressive set of credentials interviewed at some point. A lot of data are tossed around. Fancy jargon and clever punditry are used to discuss the direction of our economy.

Economics is more than this, and an economic decision-making framework is something else entirely. In a nutshell, the approach is clear and practical: *do something if the benefits outweigh the costs*. Invariably, this is what we as humans always do in the end. We seek the greatest results with the least amount of effort and cost.

This chapter is about the food we choose to eat, and why. In particular, I focus on how we, as individuals, can approach lowering obesity and improving health. These are the things people want the most, and the two things that most influence our well-being as individuals and as a society. The distinguishing feature of this chapter, what sets it apart from the gazillions of other articles and books out there, is the focus on efficiency. Efficiency, as we shall see, translates to simplicity, something that is sorely lacking in today's world of food choice. It also takes into consideration the broader picture of choice, which tends to be overlooked—with dire consequences.

Nutrition advice is often based on scientific studies conducted by researchers who are trained to think about minutiae rather than the practical challenges that people face on a day-to-day basis. Nutrition studies don't tell us how to balance trade-offs. They don't outline the costs and benefits of following a particular dietary profile. Yet all diets involve benefits, costs, and trade-offs.

"Costs" doesn't refer only to money. I am also talking about things like time, mental effort, and social costs. Some dietary lifestyles are so restrictive and demand so much planning, you spend several hours a day mapping out the next day's meals. You may miss many social events because the food served will not comply with the program. These are all costs, and they matter.

The point? Eating healthy must be made efficient. In terms of time, money, and mental energy, the benefits of being healthy must outweigh the costs. In order for an eating style to be long-run sustainable, the benefits a person reaps must exceed the costs incurred. Otherwise, people will simply give up.

Within the economic framework, there is also an auxiliary concept called diminishing marginal benefits (or increasing marginal costs, if you prefer), according to which, as you continue to exert effort on some activity, at some point, the returns from that activity will begin to decline. Thus we will have to work increasingly harder to achieve increasingly smaller benefits. Many people who exercise are intimately familiar with this concept. Usually, when people begin a workout program, they realize extraordinary benefits in the first few weeks. But over time the incremental gains will start to decline, and it will take increasingly more rigorous workouts to achieve the benefits. People in business might also recognize this concept in the form of the 80–20 rule. For example, 80 percent of your profits come from 20 percent of your products or innovations.

Drawing on the 80–20 rule analogy, it is possible you could realize 80 percent of possible benefits from only 20 percent of the possible effort by being smart about where you put your effort and focusing on the factors that bring the greatest health benefits. For most busy people juggling multiple priorities, achieving 80 percent of the gains of healthy eating is enough to avoid obesity and maintain general good health, without making "being healthy" a full-time job.[3]

Of course, there are those who want perfection. They typically strive to achieve 100 percent of the health benefits. The bad news is that, due to the law of diminishing marginal benefits, it may not make sense to try to reap 100 percent of the benefits. For example, a few hours of mindful shopping and food preparation might be sufficient to get you to the 80 percent mark. But you may have to double the amount of effort to get to the 90 percent mark. And to get to the 95 percent mark, you may have to triple or even quadruple the number of hours you spend.

If you are a world-class athlete, then these small incremental improvements can yield huge benefits over the competition. It might be worthwhile to strive for 100 percent. Indeed, we have often heard statements like, "She gave 110 percent" or "He left it all on the field" at the end of an athletic competition. But for most busy people, achieving 95 percent may make no sense at all. In fact, it may even be counterproductive in the long run. One can easily get burned out or quit a particular eating plan if it becomes too onerous. *Remember, do something only when the benefits exceed the costs.*

Another way to look at this is that we want to achieve an "optimal" balance, where the incremental benefits are as close to the incremental costs, but without exceeding it. This is one way to justify the principle of moderation. Moderation usually gets us closer to the optimal than either giving up (not trying at all) or striving for perfection. Economists typically refer to the latter two as "corner solutions," which are rarely optimal.

Is It Worth It to Try?

Doesn't it seem obvious that we would want to try to eat healthy? Well, no. Otherwise we already would be. From the viewpoint of society, it surely seems like a good idea. But there is no universal answer for everyone. Each person has a unique set of preferences and trade-offs, and this is often overlooked as we try to prescribe monolithic, top-down approaches that would work if you "just do it right." The crux of the matter is that, given a person's individual preferences, the benefits don't always outweigh the costs.

Let's take a simple example. Eating healthy over the long term will involve cutting back on processed or manufactured foods. Simple, right? Not quite. Giving up processed foods has many implicit costs. Many processed foods taste very good because the manufacturers spend lots of money to create the right balance of ingredients to enhance eating pleasure, and yes, to stimulate more eating. So giving up processed foods might very well decrease the overall food "reward" in your diet (that good feeling after you eat, let's say). This is not a trivial cost to many people. Secondly, many processed foods are convenient. They have long shelf lives and don't require much preparation. The combination of pleasure and convenience is powerful.

There are some people who are largely satisfied (or unconcerned) with their health and for whom the benefits of giving up processed foods will not exceed the costs. They should not necessarily pursue health as their main objective when choosing food. It just isn't important enough to offset the loss of that delicious, prepackaged bag of potato chips that slides so easily into the backpack. Yet most people, and these are the people I focus on from here on out, really want to be healthy. Some may have given up in the past because the costs proved to be too high, but they still want to be healthy. The diets didn't work; there was too much information and they couldn't figure it out. The only examples of healthy people they knew were stringent rule followers of extreme diets. But for this group, which is most of society, an *efficient* eating plan might tip them toward taking action.

It should be obvious that people want to lose weight and want to be healthy, but it isn't. Being healthy makes us better citizens and contributors to society. It prolongs our life, and our *quality* of life. But eating healthy is not easy in our food environment, and when the costs of being healthy are greater than the benefits, then we drift in the other direction.

So how do we create a health-oriented, *efficient* food plan? The first step is to identify the most important factors of "being healthy." Those are the ones to focus on, because with modest effort we can achieve a large return. Most people who attempt to eat healthy are often sidetracked by elaborate eating regimes that include a myriad of rules that have only small, if any, impact on health.

Society today is inundated with philosophies on how to eat healthy. It's too much. Following every recommendation makes us feel overwhelmed. It's the paradox of choice. If there are too many options, we are consumed by anxiety and become paralyzed. Another paradox, curiously, is the obesity paradox. Some people claim that higher fat mass can paradoxically lead to better health outcomes. Yet recent studies raise doubt about the existence of this paradox, and in fact a recent meta-analysis supporting the obesity paradox was found to have been, in all likelihood, confounded by other factors.[4]

Narrow the list to increase the odds of long-run success. Many recommendations have not yielded tangible benefits in scientific studies. These recommendations mostly emerged out of ideology or fears that have largely been unsubstantiated. For example, there are virtually no scientific studies published in the better academic journals that show that eating GMO-free yields tangible health benefits. GMO-free foods are also expensive and difficult to find so if one were to adopt a GMO-free diet, it would be expensive and burdensome. This is an example of a food rule that is low benefit, expensive, and time-consuming.

On the other hand, simply eating lots of fruits and vegetables, regardless of whether they are GMO, local, or organic, has been robustly shown to improve health through multiple scientific studies. Another (obvious) and very effective way to increase health is to decrease obesity. Eating to reduce obesity is what I would call a *first-order* factor in improving health.[5] First-order factors are those that yield the biggest and most robust benefits. Given that avoiding obesity confers so many health benefits, an efficient 80–20 rule approach to health might start by formulating a strategy to reduce or avoid obesity.

The Benefits and Costs of Eating to Avoid Obesity

In the grand scheme of things, there are heavy costs to being obese. Yet we don't make decisions about our weight based on the "macro perspective." We make decisions based on our individual preferences and how we balance costs and benefits. Confusion, for instance, is a cost. Frustration is a cost. There is an abundance of this to go around in today's eating environment.

I was really confused when I first started exploring new ways of eating to become healthier. I looked at the standard calorie-counting diets; low-carb, plant-based vegan-style eating; vegetarian; pescatarian; and even potato diets, along with an endless array of other types of eating philosophies if one were to google diets on the Internet. How would one go about sorting through all of these diets? I'm an academic; you would think if anyone could, I could, but advocates of just about all of these diets can cite scientific studies to support their claims.

We have to learn how to screen our diets. There are two ways to do this; the principle of parsimony (related to Occam's razor), and the concept of robustness.

The principle of parsimony roughly says that if you have multiple competing theories, go with the simplest one. There are two reasons to choose simpler dieting philosophies. First, simpler theories generally require fewer assumptions and are less susceptible to a concept called overfitting. They are more likely to perform better. Second, simpler theories are often easier to implement. This rules out many extreme diets that are excessively rule-bound, include numerous food restrictions, and have to be implemented perfectly to "work."

The concept of robustness makes two points. First, the credibility of a theory or dietary philosophy increases if there are multiple empirical studies or even an entire body of literature that supports the theory. I alluded to this earlier, but if one looks hard enough, one can usually find a single study to support almost anything. Economist Charles Wheelan makes the point that if you conduct 100 studies you are likely to get some nonsensical results out of a few studies just as a matter of pure statistical chance.[6] Given that this is the case, it is really easy for dietary dogmatists to cherry-pick studies to fit their worldview. Yet it is much more difficult to cherry-pick an entire body of work based on multiple independent studies. A theory is robust if it is supported by numerous independent studies under a multitude of different assumptions.

Second, and more practically, I consider a dietary philosophy to be robust if it does not require absolute adherence to every last detail to be effective over the long run. For instance, there are many extreme diets that seem to cause failure-to-thrive over the long run. A hallmark of these diets is that the "experts" or gurus that are ideologically bound to these diets will often blame the consumer for "not doing it right." There will be quite a few websites on the Internet devoted to how to follow these diets "correctly." This should be a warning sign.

One of the most robust and simple theories of weight loss is the "calories in versus calories out" theory. This theory is widely known and needs little explanation. It basically says that if you take in more calories than you burn, you will gain weight. If you take in fewer calories than you burn, you will lose weight. Not only is the theory simple (parsimonious) and based on well-known laws of thermodynamics, but there is a large body of evidence showing that the obesity epidemic is highly correlated with the number of calories consumed over the last several decades.[7]

Therefore, any dietary lifestyle that claims to work through some other mechanism rather than calorie-in-calorie-out should be scrutinized with skepticism, given the amount of knowledge we currently have. While there might be minor deviations in the calories model of obesity, due to the fact that certain macronutrients, such as protein, take slightly more energy to digest than carbs or fat, the general positive correlation between calories taken in and weight gain is difficult to dispute.

Now some folks may point to the carbohydrate-insulin theory (CIT) of obesity as having supplanted the calorie theory of obesity. CIT basically postulates that consuming carbohydrates elevates insulin, and insulin, in turn, promotes fat gain. CIT provides the theoretical foundation for low-carb diets, which are currently very popular. One nice thing about CIT is that it appears to be quite clear about what mechanisms should theoretically promote fat gain and can therefore be tested using controlled experiments. Thus it seems to be quite parsimonious in a scientific sense. However, it does not appear to be a very robust theory. At the same time, there is plenty of anecdotal evidence that many people have lost weight on low-carb diets. In fact, in my own personal journey in self-experimentation with various diets, a low-carb diet was quite successful in helping me shed many pounds in a relatively short period of time. How can that be if CIT has been falsified?

When people adopt low-carb diets, there is usually a big payoff at the beginning. This is usually due to water loss from the liver and muscles. While this can have a beneficial psychological effect for people trying to lose weight, it is hardly the type of weight loss we are interested in, which is fat loss. However, low-carb diets typically create indirect spontaneous reductions in calorie consumption by eliminating a major food group that humans consume. Around the world and over the last several centuries, humans consume many carb-heavy foods. So when a diet wipes out rice, sweet potatoes, potatoes, many sweet fruits, bread, and so forth, the diet becomes highly restrictive, which makes it harder to eat as many calories. My own personal experience when I was avoiding carbs was that it was difficult for me to get sufficient calories. I would often fall short of my daily caloric goals by about 300–500 per day. Also, when one drastically cuts carbs, some of those calories are replaced by protein, and it is well known that protein has a strong satiating effect, which induces a feeling of fullness so that one tends to eat even less.

There are also many studies that compare low-carb to high-carb diets in experimental settings where calories are equalized. In these settings, there is very little difference in weight loss across the diets. Ultimately, the effectiveness of low-carb diets appears to have less to do with carbs and insulin and more to do with indirect reductions in calories, which again, points to the robustness of the calories-in versus calories-out hypothesis. This should not be surprising, because many traditional diets around the world tend to be carbohydrate heavy. Yet chronic metabolic diseases were not an epidemic, as they have become in many modern industrialized societies.

Any highly restrictive diet is likely to cause weight loss through the same mechanism: the indirect reduction in calories. While highly restrictive diets are one way for people to lose weight, they tend to fail in terms of both parsimony and robustness on practical grounds.

For example, eating a high-carb vegan diet that keeps protein and fat at 10 percent of calories is extremely demanding and requires many sacrifices in terms of food preparation and enjoyment. Moreover, social events and eating at restaurants often cause anxiety because the food served is unlikely to conform to the highly restrictive diet. In addition, when I tried it, my wife complained that my low-fat cooking did not taste very good.

Another problem in high effort, nonrobust diets is all the rules, most of which may have little to no impact on health or weight. For example, the most restrictive low-carb diets suggest that people avoid sweet fruit, such as apples and bananas, due to their high sugar and carb content. But avoiding these and other tasty fruits, such as watermelon and grapes, has very little to no impact on weight loss based on numerous studies. And anecdotally, one would be hard-pressed to find an overweight fruitarian (someone who eats mostly fruit).

We want to identify the features of diets that have the highest impact on weight loss. And we want to get rid of the features that have little to no impact. This is sorting out the first-order rules from the others. For most of human history, humans ate out of biological necessity. We evolved as a species to seek out calories rather than to avoid them. Over the last half century or so, all of this seems to have taken a 180-degree turn because now we are encouraged by health authorities to *limit* calories rather than to seek out more calories like our ancestors did. If we took a step back and thought about this, it seems almost perverse. But it is not all that shocking when we consider that we live in an environment characterized by abundance and with food companies innovating new products to increase our eating enjoyment. Our problem has now switched from seeking out calories to finding ways to exercise moderation in our consumption.

According to research, the most robust factors in predicting weight gain are food palatability (does it taste good?), calorie density (how many calories per unit of food?), and satiety (does it reduce hunger?). Food palatability has to do with how good food tastes. There are numerous studies showing that food palatability is a robust factor in predicting obesity. In short, foods that light up pleasure centers in our brain can be a robust predictor of caloric intake. You might say that food is one of the great pleasures in life, and it would be a dull life to lose weight by eating bland food. Good point. I agree. However, what we are concerned with here is not just good-tasting food, but rather, unnaturally good-tasting food, such as manufactured or processed foods. When obesity researchers talk about a "change in food environment," they are usually referring to the increased availability of modern manufactured or processed food products.

To be clear, I have nothing against processed foods. I don't consider them "evil" like some foodies do. Our modern economy consists of many enterprising firms innovating and creating new products that enrich the consumer landscape, whether it be new technology and gadgets, cars, TVs, music, movies, and the like. Manufactured foods are just another set of "consumer goods" emerging from this product innovation culture. Consumer products are engineered to increase sales, and the only way to do that is to create products that consumers want. But part of how manufacturers increase palatability is to create food products with "unnatural" combinations of fat and sugar, both of which taste good in isolation but can become spectacular when properly combined.

While I consider having unusually good-tasting foods to be no more inherently evil than, say, having the latest and greatest smartphone with dozens of capabilities and instant access to information, it does mean that we face different sets of trade-offs today than in the past. Just as technological innovation has reduced face-to-face interactions, increased Internet addiction, and increased certain risks for children in an online environment, modern manufactured foods have made it easier to overeat and overconsume. For people who are interested in becoming healthy, it means that they have to be perfectly clear about the trade-offs. Is being healthy worth the exercise in self-control and moderation, or is their indulgence in a richer food product worth more?

The other important factor in driving obesity is calorie density. Calorie density is also related to satiety because foods that tend to have low calorie density also tend to be more filling. Apples tend to be more satiating than croissants. Anybody who has tried to eat 300 calories of apples at one time knows how difficult this is. Yet it is very easy to eat 300 calories of donuts or flavored tortilla chips in one sitting. Low-satiety foods promote overeating because one tends to feel less full after eating them.

If high palatability, calorie density, and low satiety are first-order factors in inducing weight gain, how do we translate this into a specific eating strategy? Consume nonmanufactured foods that are lower in palatability and caloric density while being high in satiation. Nonmanufactured foods, such as fruits, vegetables, beans, meat, and seafood, tend to be water dense, high in fiber, or high in protein and have exactly the factors that research has shown to increase satiety. Sugar-sweetened beverages deserve special mention. They taste great, are calorie dense, and are low on the satiation spectrum. A systematic literature review conducted in 2006 suggests that the consumption of sugar-sweetened drinks, particularly soda, is highly correlated with higher body fat. The key reason is that sugar-sweetened beverages provide nontrivial amounts of calories while providing low satiation. For people interested in losing weight, avoiding drinking one's calories can reduce spontaneous calorie consumption.

The Punchline

Here it is: *If you are making food choices to be healthy, you get the biggest bang for your buck by focusing on foods that will allow you to maintain a healthy weight (i.e., avoid obesity).* The following strategy is one that can maximize benefits while minimizing costs for the average person (lives a sedentary life, works a normal job, is busy):

1. Eat a diet with as many minimally processed, nonmanufactured foods as possible.

2. Eat calories rather than drink calories.
3. Make sure to include protein in your diet to increase the feeling of fullness and to preserve lean mass.

Of course, in practice, the line between processed foods and nonprocessed (whole foods if you prefer) can be blurry. For example, is bread a processed food? Many of my colleagues would also tell you that there are no clear definitions of processed or whole foods, and I won't attempt to offer one. Others will also tell you that some processed foods might even be healthy for you. And I agree.

However, this is about keeping it simple. We are trying to follow the 80–20 rule to maximize benefits while keeping costs low. Simplicity is extremely important for an eating approach to be long-run sustainable. For the average person, we need the economic approach of keeping costs low. And if someone has doubts about whether a food product is processed or not, my suggestion is that they go ahead and eat it rather than become neurotic about it. Obsessing over minutiae and food perfectionism can promote eating disorders, which can lead to a whole host of other health issues.

Personally? I try to get around 3,000 calories a day to maintain my body weight. About 1,000 of my daily calories come from fruit. Another 1,000 come from steamed potatoes, rice, eggs, or unprocessed chicken or seafood, depending on the day. Then I eat whatever I want with my remaining calories, including ice cream and gummy bears (a real weak spot for me). Even so I've had no problems with weight gain, and all of my health markers are in their optimal ranges according to my doctor.

Also for the sake of simplicity, I do not suggest that people worry about specific targets for macro- or micronutrients. As Dr. David Katz, the director of the Yale University Prevention Research Center, was quoted as saying, "If you focus on real food (food from nature), nutrients tend to take care of themselves!"[8] Essentially, Dr. Katz is saying that if people eat a variety of minimally processed foods, they can stop focusing on individual ingredients or nutrients because the foods themselves will provide the right dosages in the right combinations.

In the name of keeping it simple, and the costs low, from the perspective of avoiding obesity it is not necessary to eat GMO-free, organic, or local or to follow whatever is the current fad among foodies. If you have other reasons to eat according to these rules or have personal convictions, that's fine. For example, if you want to support your local farmer, that is laudable. But from the perspective of weight loss, these will not generate first-order effects.

Plus, GMO-free, locally sourced, organic foods are expensive. For the truly budget-constrained, telling people they "should" eat these foods sends the wrong message to those who have stringent budgets or live paycheck to paycheck. It is much more important to get your fruits and veggies everyday rather than avoid them because they don't check a list of boxes that create only second-order benefits for managing obesity. Conventional whole foods can be quite budget friendly per calorie and an improvement over cheap processed foods. The other day, I bought 1,000 calories worth of bananas for $3.13. Of course, if you have money to burn, then I encourage you to buy whatever you want, and in some cases you may even realize some additional health benefits. But these are usually going to be second-order health benefits from chasing the remaining 20 percent of benefits rather than the first 80 percent from following the 80–20 rule. For the same reason, I have focused on avoiding obesity as the primary health goal in this essay. While there are certainly other health markers aside from being overweight, losing weight if one is obese tends to yield the largest health benefits so that it is consistent with the 80–20 rule.

One of the best pieces of advice I have read is by nutrition researcher and educator Alan Aragon. Look at your target weight or end goal. Then adopt a food lifestyle and exercise regime that is consistent with that end goal. In other words, act like the person that you want to be. Obtaining health is easier when lifestyle changes are permanent. Ditch the crash diet. A study by a group of researchers reinforces this; weight loss maintenance becomes easier after individuals have maintained weight loss for two to five years.[9]

No Such Thing as a Free Lunch

Everything is a balance between cost and benefit. What price will we pay for the benefit? What will we sacrifice to get our desired object? What I outlined here is a personal approach to tackle health in your own life, because diet *is* deeply personal and frankly can't be foisted on someone else. Understanding an economics decision-making framework is a tool to use to cut through all the noise. Then you can make the right changes for yourself.

Today, there is a lot of noise out there. Too much noise. This translates to increased "costs" in pursuing good health. The path to a practical eating plan is a veritable quagmire of frustration, confusion, temptation, misunderstandings, and misconceptions. Most of these are wrought by forces outside of our control. Spend two minutes flipping through a magazine or watching TV, and chances are you will see an advertisement for something about your diet and health, whether subliminal or smack in your face. This is a problem, and this is a grave challenge we face as a global community. Today, the scales aren't tipped in the favor of good health. By understanding the costs/benefits equation by which we all live, and eat, perhaps we can work to realign the balance not just in our own lives, but on a broader scale.

Chapter 10

Social License to Operate

Nicole J. Olynk Widmar

Cultural norms vary across the globe. Can we work across our differences?

Back in 2012, McDonald's announced it was starting to phase out pork from farmers using gestation stalls. Gestation stalls are a housing system for pigs. This type of housing system tends to reduce aggressive behavior as pigs can be rather belligerent creatures, particularly toward one another. An alternative to gestation stalls is group housing. The question as to which is right isn't obvious; research has found that both gestation stalls and group housing have unique benefits and drawbacks.

Gestation crates were and are legal, and there aren't any government sanctions against them.[1] Their use did and still does comply with food and health standards. So why then did McDonald's announce it was going to phase out pork from farmers using this system? The decision would be costly and, given the scope of McDonald's and the amount of pork they purchase, the move carries major ramifications for farmers and production systems.

In a nutshell, the decision was driven by pressure from the public. Patrons of McDonald's exercised their discretion as consumers. Government regulations and rules weren't enough. They communicated that they would not purchase a certain product for reasons *beyond* those generally mandated. Consumers didn't like the use of gestation stalls, perceiving them to be bad for the welfare of the pigs, and McDonald's felt that pressure. So they acted.

Today, there is an abundance of examples in which food products are phased in or out of favor based on production system attributes, such as, in this case, an animal housing system. These attributes are beyond what is governed by regulation or traditional legal structures. We, the consumers, wield our purchasing power to influence the business decisions of major agglomerations. We make it clear what is, and is not, acceptable *beyond* what is required by law.

We are all familiar with the concept of a license. Some body, usually the government, grants us approval to do something. Probably top of mind is a driver's license. We pass a written test and a driving test and prove our identity to the government, they issue us a license to drive. We are then legally sanctioned to be out on the roads. Doctors, lawyers, and a range of other professionals pass a regulated exam to receive a license to practice.

In addition to legal licenses there are also social licenses. Members of a society grant, or revoke, the privileges of a business, individual, or institution to exist or operate within said society. This concept is commonly known as a social license to operate. McDonald's can't stay in business without an actively engaged and happy group of Sausage McMuffin lovers. They won't generate the profit they need to continue operating, even though they may be meeting all the legal standards and regulations to remain in business.

The so-called social licenses to operate have been commonly discussed in recent media around issues ranging from mining[2] to food safety.[3] The license is granted and revoked based on concepts, values, tools, and practices that represent a way of viewing reality for an industry and its stakeholders. If the concepts, values, tools, and practices of a company aren't acceptable to an influential enough swath of society, then their social license to operate is at risk. If we state loudly enough, whether with our words or with our purchasing power, that what they're doing is not okay, and if enough people get behind us, we wield the power to grant, or revoke, a company's social license to operate.

At one time, finances were the primary driver of the decisions made by a business or organization. Today, however, social license to operate has become a vision of how to move forward with new projects. It adds a new dimension of acceptance. For example, if a community doesn't support the development of a mine, it doesn't matter how high the commodity prices are. There is no guarantee a positive production or development decision will be made.[4] If production is morally or ethically unacceptable to society, production either won't happen or will struggle mightily.

Our ability to influence decisions and control social licenses is both a right and a privilege. Yet inherent in this is a frequently overlooked and grave challenge. We don't all see things the same way. For instance, we have different views of animals and we have different views of various species of animals. Different cultures view food production issues differently, and even different *people* view food production issues differently. Additionally, we often run the risk of becoming swept away in moral outrage before we have appreciated the full picture.

Using our power to revoke and grant social licenses to operate carries consequences. We can do great things with it. We can also do harm. For instance, failure to understand and appreciate differences in opinion and the nuances of a situation has the potential not just to challenge and disrupt trade relations but also to fuel social discontentment. Interrupted trade and social discontent are serious things indeed when it comes to food security. And thus, our capacity as a society to grant social licenses to operate faces a unique challenge. Can we do so responsibly and with an attitude of tolerance toward different preferences? Can we support, or at least grant social licenses, to organizations (or industries) moving the needle toward positive outcomes deemed advantageous or improvements over the status quo? Or will we become so stubborn in the righteousness of our beliefs that we disrupt, interrupt, and diminish our capacity to feed the world?

It Is More Than Just Food

Caviar and champagne are the food and drink of the wealthy. Orange Crush and white bread smeared in generic grape jelly are sustenance of the poor. Craft beer accompanied by quinoa and tofu sprinkled with cilantro and a smattering of sun-dried tomatoes drizzled in extra virgin olive oil with a dusting of fresh ground white pepper is, naturally, manna of the hipster.

Food represents status symbols, whether we are attuned to it or not. By a glance, whether subconsciously or with full awareness of our critique, we garner facts and details about an individual's socioeconomic status. These status symbols are not universal; they vary across cultures and regions. What denotes wealth and prosperity in one place may do the opposite in another. What symbolizes prestige and status in one country may be meaningless in the next.

Then, of course, there are the rituals, celebrations, and religious beliefs. Turkey for Thanksgiving. Cake and ice cream on your birthday. Brats and burgers on the Fourth of July. Pancakes on a Saturday morning. A big roast for Sunday dinner.

Already, food has become something far more than a means of sustenance. It has become an indicator of wealth and poverty, of culture and belief systems. Loaded with far more than calories, it is an entire sociological analysis of self.

This applies even beyond the food itself to include how we consume it. Lack of knowledge on how to eat "properly" is a universal sign of "outsider status."[5] Consider tea services and table settings. These intricate settings and ceremonies readily reveal the uninitiated. We may use the improper fork, partake from the wrong water glass and bread plate, or botch the timing of the meal. Even knowledge of food, without a scrap of food in sight, is often perceived as a signal of prestige or class. For example, a knowledge of foreign foods may signal one's cosmopolitanism.

And so, it becomes clear how incredibly fraught food is with symbolism of status and identity. It leaps far beyond a simple source of energy for keeping ourselves moving throughout the day to create a stamp of self. It reveals our morals and ethics, our religion and wealth, our health status, our nationality, and our education and experiences.

We have a clear interest in our food even when there's no need for us to be involved with its production. Given this attachment in terms of tradition, ritual, culture, it's no wonder we have an innate desire to understand the processes that contribute to its existence. It's no surprise that support of life movements, such as slow food, raw food, and local food, have gained traction. And even without a "branded" or recognized movement, plenty of people are deeply devoted to personal systems, such as the home-cooked meal, made-from-scratch baking, home canning and preserving, and even vacations or travel devoted to cooking and eating.

Much of our direct involvement in food production has diminished. Where once most people worked in agriculture and food production, now only a slim portion of the population does so. Yet our interest in food has not diminished in the slightest. For many, this interest has been reallocated to understanding and, in some cases, influencing, how food is produced and brought to the marketplace.

The Subjective Animal

An animal can be a pet. It can also be a nuisance, causing property damage or stealing food from the birdfeeder. Perhaps it is an athlete of sorts, performing at sporting events, such as horse racing. Maybe it is a form of entertainment, captivating audiences at zoos or wildlife-watching locations.

Animals may have a job, possibly as a service animal, helping to guide the blind, for instance. Or what about military and police dogs? Search and rescue animals? Then there are the less obvious roles of animals, the ones that are seldom witnessed. The existence of the polar bear in its natural habitat, while hardly ever witnessed, is pleasing to many of us around the world and has become something of a symbol.

Animals carry social importance in many cultures, including that of the United States. For people who don't work in agriculture, farm animals are often far more relatable than images from crop systems. For example, the farm scene in your mind's eye often involves a black and white spotted cow, a pink pig, and a white chicken. Consider asking a young child to draw a cow or a chicken. Most children could draw something recognizable as a chicken; the chicken would likely be a circular figure and likely colored white. Farm animals are in some sense universal symbols of a perfect farm scene with green grass, blue sky (dotted with perfectly puffy clouds), and the cow, pig, and chicken in the barnyard. Seldom does the romantic notion of production agriculture conjure up visions of the perfect soybean. Ask the same child to draw a soybean and see what you get. Your guess would be as good as mine.

Animals are with us from the beginning, with stuffed animals or in cartoons as children. Many kids possess at least one stuffed animal, but very few possess the seldom-seen stuffed corn kernel. Animals are presented as friends, consider the Cat in the Hat, who has taken decades worth of children on many an adventure in books and television. Then there's Winnie-the-Pooh, who has been a friend to children since 1926. Pooh Bear, as he is known to his friends (of which I am one), is prominently featured in books, plush toys, clothing, and videos for children. As perhaps the most indisputable evidence of his fame, the rights to Winnie-the-Pooh were licensed to Walt Disney Productions.

Given all this, it is easy to see how animals play many roles in our lives and society. While most, if not all, the aforementioned roles are socially acceptable to those of us in the United States, this is not the case everywhere. Some cultures find these appalling, others lovely, and some might land in between.

Yet there is no role for animals in the lives of humans that elicits more discussion and debate than when animals are consumed as food. Some individuals would stridently insist that animals should never be eaten under any circumstances. For others, doing so would betray their religion. Plenty of people eat hamburgers, chicken nuggets, and all kinds of animal protein without a second thought, untroubled by the source of their food or the treatment it received across its life. Some may eat them, but only if they were raised in a particular way.

What's more, it might be okay to eat this species of animal, but not that one. For example, eating dogs and horses is completely normal in some places, and extremely taboo in others. Consider that barbeque invitation from your new neighbors. You whip up your potato salad, walk across the street on a warm and sunny Sunday afternoon, and are then served horse steaks. How would that go over?

It probably depends on how you classify a horse. Is it a pet? A farm animal? A source of food? A romantic throwback to the yearnings of your teenage years?

Whether we find that horse steak palatable depends on a few things. For starters, it often rests on the classification of the species. For example, an individual who classifies an animal as livestock might be expected to be more accepting of eating that animal. In contrast, classifying an animal, say a dog or cat, as a pet may create the expectation that the animal provides companionship or perhaps household security, but is not necessarily food for human beings.

Other factors influencing our feelings can be perceived beauty and personal emotional attachment. Even more, the acceptability of eating a species (e.g., beef cows) might be okay, but there is no way we're going to eat Betsy, the sweet, brown-eyed pet cow from our neighbor's farm. These things can be very situational as well. I have a friend who is an avid bird-watcher. During a trip to China, her guide excitedly pointed out a common but beautiful shore bird seen almost worldwide in abundance. To the guide, it was a rare sighting because, during the Chinese famine of the late 1950s and early 1960s, this bird was nearly wiped out in China because it was used for food. To my friend, the thought of eating it was abhorrent, but she probably would have partaken if she were starving.

Research has sought to understand how people categorize various animal species to aid in understanding how people view their roles in society. For example, a study of US residents found that over 90 percent of people classified dogs and cats as pets, whereas only 58 percent classified rabbits as pets and 55 percent classified horses as pets. In the same study, 27 percent of respondents classified horses as livestock animals. In contrast, over 80 percent of respondents classified beef cows, dairy cows, pigs, chickens, and turkeys as livestock animals.[6]

Animals' roles in society are complicated and fraught with emotional attachments. People are less open to eating animals that they classify as pets. No surprise, right? So if someone perceives their horse (or pony) as a pet, the horse steaks featured at the neighbor's barbeque may be hard, or even impossible, to accept. On the other hand, if horses are labeled in one's mind as farm animals, eating them may be just fine.

Wielding Influence

Imagine we are in the marketplace. It could be our favorite neighborhood supermarket or restaurant. We are perusing the aisles, or reading the menu, in search of dinner. We express ourselves through our purchases, or we "vote" with our money. If an item sells quickly and in great quantity, the maker of the item is incentivized to make more. The market has provided the signal that "people like this product (at the stated price) and are buying it, so make more so you can sell more." On the other hand, if no one buys a product, it sends the signal that "nobody is willing to pay for your stuff." Over time, the maker of the lesser-liked product either adapts to make it more desirable or they stop making it entirely.

When it comes to our food, our "votes" revolve around many factors, some of which are social concerns. Social concerns today revolve around factors such as the environment, food safety, animal welfare, the use of antimicrobials (antibiotics) and the potential impacts on human health and medicine, worker/labor well-being on farms (and in processing facilities), and impacts of agricultural trade on other economies and populations. Of course, this is just a sample; there are many more.

These social concerns have wrought change. For instance, modern livestock operations are being asked (actually forced) to comply with expectations from newly vocal stakeholders who were once less directly influential in shaping on-farm practices. These newly vocal stakeholders would be us: the food shopper, restaurant-goer, and all-around consumer of food.

In order to remain in business today, livestock producers can't just meet industry standards, regulations, and health and safety inspections. They can't just comply with various levels of zoning and siting regulations, including environmental impact assessments, water collection facilities, and building standards. It is not enough to be in compliance with every formal system, regulation, buyer standard, environmental agency, and zoning board in the locality of operation.

They must also meet the regulations of their buyers. As we saw in the case of McDonald's, a huge purchaser of pork, in response to the preferences of *their* customers, they were going to phase out pigs raised in systems using gestation stalls. So a livestock business hoping to sell to McDonald's must be in compliance with both the formal system and the demands of their customer (McDonald's).

Purchasing food isn't the only way to influence companies. People have found many other ways to wield influence, through litigation, public relations campaigns, social media buzz, the media, community organizations, and more. The growth of "politics by other means—politics practiced through the market," has allowed interest groups to pursue political changes through the market system rather than legislative channels as was traditionally done.[7] One key example of such movement is the disadoption of recombinant bovine somatotropin (rBST), also known as recombinant bovine growth hormone (rBGH) in milk production. Due to a set of actions and reactions set into place by retailers' movement away from milk produced with rBST, milk producers were "forced" to adjust their production practices to meet the demand of their customer—the retailer. In this way, politics practiced through the market have led dairy producers to move away from the use of rBST, although no regulation or legal action has been taken to eliminate the use of the technology.[8]

My Way or the Highway?

Most of us in the developed world today are savvy consumers of social media. We read, see, repost, and discuss more socially minded issues related to food production than in the past. As a result, for many of us today our food choices convey not only status, culture, and education but also our social values.

The food industry and production agriculture are experiencing increased scrutiny from consumer groups demanding "more" in terms of attributes from their purchases. There is a growing concern over human labor, and people in the United States and other Western nations are becoming progressively more concerned about the general care and well-being of livestock animals, such as the concern over sows and gestation stalls.

A result of this growing interest in social responsibility is that corporate social responsibility concepts have developed into top priorities for many businesses. As such, earning a social license to operate has become top of mind. For many businesses, this means showing their consumers that they are practicing accountability, credibility, flexibility, and capacity.[9] McDonald's sought to demonstrate to their customers that they could be both flexible and accountable when they decided to phase out pork raised in gestation stalls.

This is all great, right? In some cases it is. In some cases it isn't. An unwillingness to accept variations in perceptions of animals, for instance, tends to be inherent, as it is a highly contentious issue that people are passionate about. Wealthy consumers may be blind to the fact that their demands will raise the price of food and, as a result, drastically raise the financial stress a poor person might feel. Or what if, in some cases, people are acting off misinformation or a lack of context? They may not realize that supporting a certain cause or passing a certain legislation may destroy a community's economy or lead to unemployment for certain producers.

For some, this may not matter. Financial stress on the poor due to raised prices or the unemployment of a segment of society isn't as important as their moral beliefs toward food. But an unwillingness to accept alternative views of animals and their uses can challenge the food system. Those who disagree with using specific species for food may also wish to block the ability of others to do so.

Take the contention around animals and their treatment. It is not inconsequential. As perceptions of animals differ around the globe, it can disrupt worldwide food production. It spills over into trade negotiations and, if it is contentious enough, it can even impact the trade of non-animal-derived products. Trade is, as we saw in an earlier chapter, complicated, delicate, and extremely important for food security.

Consider, also, the potential for subtler social implications. Let's say that one country allows dogs to be raised for slaughter. Their consumption of dog meat taints your entire opinion of the residents of that country, particularly if you learn about their eating habits while scrolling through Facebook with your poodle FiFi by your side.

Our positions and stances on food production issues can yield tremendously positive advances. The challenge is to maintain balance. If we move beyond the pale of tolerance and understanding, such that our passionate stance on any one issue doesn't allow for dissenters or cultural differences, we can revoke social licenses to operate in ways that are ultimately damaging to ourselves and our communities. In a world dependent upon peace and trade for food security, we confront the challenge of carefully and thoughtfully selecting which social licenses are granted, which are revoked, and the way in which this is done.

Chapter 11

The Information Hinge

Jessica Eise

Will communication make, or break, a food-secure future?

My grandmother was born in 1917 in Madison, South Dakota. She was a child of the Roaring Twenties, and her young life was pleasant. Grandma recounted early memories of dancing, laughter, and plenty. Yet everything changed when the Great Depression struck in 1929. It brought economic despair and a Dust Bowl that wreaked havoc across the Great Plains of the United States. As a result, my grandmother's life was changed forever. Her father lost his job and moved away in search of work. Her mother turned their family home into a boarding house so they could earn enough to scrape by.

In this new reality, waste had become a deadly sin. Not a *single thing* was wasted during those tight times. Not clothing, not furniture, and most certainly not food. This painful lesson was one she carried for the rest of her life. She taught my mother to never, ever waste food, and then my mother taught me. It is a legacy passed down across the generations, born from the suffering of a poor teenage girl in terrible economic straits during the Great Depression.

Growing up, I just assumed, like many of us do, that what I had been taught was normal. Everyone thought that way about food waste, right? I believed that until I went to college. In shock, I watched people throw food in the garbage without a second thought. My friend would buy strawberries, forget about them in the fridge for a day too long, and then dump the whole container in the trash.

In the previous chapter, my colleague Nicole J. Olynk Widmar discusses the matter of a social license to operate. We, the people, allow for things to happen in our society. We also disallow them. To do so, we yield our social influence. As Nicole so neatly illustrated, we exercise this power often in matters surrounding food and agricultural. We purchase one item over another, post on social media, sign petitions, vote, and choose to throw food away or preserve it (some of us have grandmothers who were raised during the Great Depression, some of us don't).

Yet our social influence extends well beyond just the matter of food waste. It comes into play with *all* the previous challenges addressed in this book. Decisions regarding population policies, water use, land development, climate change mitigation, technological innovation, the adoption of food systems, international trade, health, and food loss and waste all come down to how we individually wield our power as members of society. Our perceptions of what is acceptable, or unacceptable, can lead to mighty forces of change indeed.

How do we, then, make these decisions? Contrary to what we may think (or wish), the right social course of action is not predetermined. This is so clearly portrayed in the case of my inherited disinclination to waste food versus the casual disregard of many of my peers in college. Not everyone approaches issues the same way. It is not inevitable that we, as a society, sway one direction rather than another. A force comes into play that influences whether we will make this choice over that, whether we will act or not, or whether we will become outraged or just simply bored.

This mighty force is communication. It is the endless flow of information between us and the greater world. It is the conversations we engage in, the classes we take, the articles we read, the social media with which we engage, the commercials we watch, the organizations we join or don't join, the celebrities and journalists to whom we listen, and the lessons we learn from our grandparents.

Communication informs our decisions. It informs how we choose to act and how we wield our power as consumers and members of a broader society. But its strength and power can be offset by its fickleness and imperfection. Effective communication is hard, and communication around food and agriculture is not yet as good as we need it to be. Ensuring that we have a reliable and trusted flow of information around the challenges we face to feed the world is, in fact, a challenge in itself.

The Communication Evolution

Back when my grandmother was born, about half the entire US population worked in agriculture. With so many people in this sector, chances were high that nearly everyone knew someone who farmed. Growing up in South Dakota, my grandmother certainly engaged in more than a few conversations about weather with local farmers living around Rapid City, and she went to school with farm kids. She would have known when it was planting season, as well as when it was time for the harvest. She surely shook the calloused hands of farmer neighbors at church while looking into eyes set in sun-browned faces with weather-beaten brows. She *knew* the people from whom she got her food and was familiar with the basic concept of agriculture, just by association with her neighbors, friends, and relatives.

But by the time my mother was born in the late 1950s, the number of farmers had radically declined. Technological innovation, as explained in chapter 5 by Uris Baldos, led to an unprecedented rise in efficiency on the farm. Fewer farmers could feed more people—as of 1960, one American farmer could feed 46 people.

What is really astonishing, though, is the leap that took place between my mother's birth and my own, as illustrated in chapter 6. By the 1980s, when I was born, one American farmer could feed 115 people. And today? One single farmer can feed 155 people—more than triple the number of people a farmer could feed at my mother's birth. With farmers being so much more efficient, we need fewer farmers, and so today less than 2 percent of the American population works in agriculture.[1]

This distance breeds an unfamiliarity with agriculture that our ancestors did not share. When my grandmother was a child and half the population still worked in this sector, she didn't feel mistrust fomented by distance and unfamiliarity. Agriculture was there, in her face, and such a way of life that I doubt she or others in her community thought twice about it. Of course, there were some people back then who lived in large cities and didn't have this type of exposure, but their numbers can't be compared to those today. The twentieth century boasted a massive rural to urban migration within the United States, as chapter 1 details. In 1900, the majority of people lived in rural areas. Today only one-fifth of our population is rural.

So how can we expect to know farmers, or feel connected to the source of our food, with so few of them out there? It is a serious communication challenge. However, this challenge arose just as two other phenomena did—two phenomena that, at first glance, ought to have brokered the distance between the general public and agriculture: the advent of the Internet, and the rise of the food movement.

The Internet has ushered in a new era of how we receive and exchange data. Journalists were traditionally known as the gatekeepers to what we would see. Their role, as defined by society, was to carefully sift through the events of the time, select newsworthy items, and strive to create a balanced, straightforward portrayal of the facts. Most information that was published (although not all, as there have been dubious publications since the dawn of publishing) was held to a certain standard. Despite budget cuts in newsrooms,[2] the same largely holds true today—for journalists. Yet for bloggers, celebrities, talk show hosts, radio stars, YouTube phenomena, and anyone with a social media account, the only barrier to entry is an Internet connection.

We have all heard the tongue-in-cheek joke, "If it's on the Internet, then it must be true!" The humor is, of course, that anyone can post anything they want on the Internet at any time. It can be as fantastical, or as authentic, as any one user pleases. Apart from credible news sites, there is precious little information curation on the Internet. It is true that libraries, educational institutions, and unique platforms like Wikipedia all attempt to curate information—but only that which they directly generate. The rate of "information" generation has outpaced the ability of any entity to validate or invalidate what is being created. Companies, social media divas, and the litany of other voices out there are not held accountable to any of those standards. They can portray any issue they want to their advantage, or to the advantage of whomever they please.

Coinciding with the rise of the Internet has been the rise of the food movement. We can think about movements as trending; they come and go. The 1960s saw the rise of the Civil Rights, counterculture, second-wave feminist, and other movements. In the 1970s, a mainstream environmental movement was born. The 1980s saw the AIDS movement. In regard to our topic at hand, over the past two decades we have seen momentum slowly growing and subsequently exploding around the food movement. Topics such as GMOs, pesticides, organics, animal rights, local food, and others are trending throughout today's society just as topics of other movements have trended in the past.

The food movement, in a nutshell, denotes the many people across the nation who are becoming or already are passionately engaged or interested in their food and where it comes from. They have an interest in how the food-supply chain works, and how the food we eat gets from farm to table. The food movement looks at and attempts to critique or influence the way we eat, and in many cases has contributed to positive gains across our food system. It has also done a fabulous job of getting people engaged in and interested in an area that, prior to the food movement, most found downright boring.

What is surprising, however, is that the food movement and the rise of the Internet didn't create a smooth bridge of exchange between the 98 percent of the public who doesn't work in food and agriculture and the 2 percent who does. The food movement generated the interest in getting the information, and the Internet could make information available. Yet we face much contention, misinformation, and mistrust.

Ultimately, the problem was that it produced groups of people talking past one another. Coming from such different backgrounds and cultures, it was difficult to relate and understand one another. Rural America, where most of our farming takes place, faces different challenges than urban America, and it has a different culture. What's more, the nature of the Internet makes it only too easy to hear those on "our side."

Add to this that the rise of the food movement, which sparked a keen interest among society about how our food is grown, took the agricultural sector by surprise. Most were largely unprepared for a public voraciously seeking information on issues that had previously been seen as too boring or irrelevant to consider. They were slow to engage, and slow to take people's concerns seriously.

Contention and Misinformation

To be clear, contentious communication around agricultural issues is not new. Robert Bakewell, now famous as the pioneer of selective breeding practices during the British Agricultural Revolution in the eighteenth century, faced considerable public controversy. His idea, now commonly employed and unquestioned, was to select animals with desirable traits (biggest, healthiest, fastest growing, etc.) and breed them. After 15 years of this practice, his average lamb increased in size from 22 pounds to 38–40 pounds. For his efforts, Bakewell was denounced from religious pulpits all over England for "playing God."

Although we live in a time far different from Bakewell's eighteenth-century England, human responses to the unknown remain the same. When it comes to food, we tend to shy away from that which we do not know.[3] Poor communication places us in an even more vulnerable position, where it is only too easy to misinterpret reality or succumb to fear-inducing snippets of "information." Our decisions are significantly affected by framing manipulation.[4]

It is not uncommon to pour our energy into concerns that aren't as critical as they are commonly portrayed to be, such as Bakewell's sheep back in the eighteenth century. For example, many fear that GMOs aren't safe to eat.[5] They are.[6] This doesn't, however, imply that a person should support them. There are a host of other reasons why a person may not support GMOs, from pesticide resistance to concerns about intellectual property rights. But studies have shown that genetic modification of crops, in and of itself, does not damage health. When an issue is misrepresented, we can fail to work together to solve critical challenges because our attention lies elsewhere.

What brings this about? It is not the religious pulpits, from which Bakewell was decried. Rather, fake news, "alternative facts," 24/7 pundits, a raging 24-hour news cycle, corporate marketing, advertisements, and the radical change in information curation have wrought a new reality. In this environment, neutral, or "boring," sources of information are often overwhelmed by glamorous, entertaining, or outrageous messages that strike an emotional chord and get us roused up. It is only too easy to pick up these sound bites, images, or solitary articles that boil an issue down to a single sentence or sweeping generalization. If you do a simple Google search for GMOs and click the images tab, there are a number of manipulated images of apples, green peppers, potatoes, and tomatoes with syringes sticking out of them. The syringes are full of unnaturally bright, alarming liquid. To a scientist, this is so far from the truth it is laughable. Yet to many others, with little background in the field, it provokes a gut response.[7] These clickbait images are meant to provoke horror, and they do, even though they are not accurate. But how can people without a background in the field know that?

Examples of poor communication sweep the agricultural sector. The case of neonicotinoids illustrates yet another instance. In a two-year, multistate study across the Midwest, scientists examined if this soybean seed treatment (neonicotinoid thiamethoxam), used to manage aphids (insects that can reduce soybean yield), actually worked. To put it in perspective, neonicotinoids are applied to more than 80 percent of corn crops in the United States and well over half of the soybeans grown in North America. Their study found that neonicotinoids did not really have any benefit for managing the aphids.[8] They also pose a risk to honeybees.[9] The companies that supply this chemical have no incentive to undertake this research, let alone communicate these messages to farmers, because they will lose massive amounts of business. As a result, farmers are then presented with conflicting messages—one from scientists, the other from the company they buy from. What is the right thing to do for their crops? What is the right thing to do for the environment? The way in which this scientific discovery is communicated will have massive ramifications.

These various waves of information from companies, advocates, politicians, the media, scientists, and celebrities all compete for airtime. And depending on the source, some aren't always concerned with presenting the whole truth or speaking from a place of thoughtfulness or compassion. This situation is perpetuated as the new ways in which we receive information make it easier, quicker, and simpler for the outrageous to spread. Social media algorithms, such as the ones deployed by Facebook, create an echo chamber of our own opinions. They show us what we "want" to see. In other words, they select posts to appear in our newsfeed that reflect similar items we have posted ourselves or that we have "liked" in the past. This new culture means that, in many cases, we pick our side before we've even consulted the facts. We diminish our space to learn and communicate. We gather allies who reinforce our position, thereby insulating ourselves further from a more concerted effort to become fully informed.

This reinforces an "us versus them" mentality. As a result, we often take the easy but illogical approach of labeling those on the other side of the debate as unworthy of being believed. Farmers and corporations become greedy entities hell-bent on polluting our environment and poisoning our food, whereas consumers are labeled as idiots who are so selfish they are willing to allow the whole world to starve as long as they have access to what they want. This is dehumanizing. It does not promote progress, and it runs contrary to the overarching reality that people on both sides of any discussion deserve respect and probably share more in common than they realize.

With the prevalence of the aforementioned and other marketing techniques, clickbait practices and blogging tactics, fear, outrage, and mistrust have been used as effective marketing tools, creating a segment of the population that is nervous and fearful toward the food and agricultural industry, and vice versa. Compounding this has been a traditional lack of transparency and unwillingness to demonstrate compromise on behalf of the food and agricultural sector, all of which are changing rapidly today.[10] Yet this has damaged trust between different groups of stakeholders and impeded effective communication around issues of critical importance in our quest to feed the world.

Finding Solutions

We are all well equipped to improve our communication environment around food and agriculture, both on a personal level and on a larger scale. The goals of both efforts are the same: communication that begets more reliable information with which to make decisions and communication that aids us in working with one another. Overcoming the challenges described in this book will require a broad, concerted effort worldwide.

Starting small, there are six ways we can each personally take responsibility for improving our communication around food and agriculture. They require only a willingness to put in the effort, and they are something we can all engage in and remind ourselves to do.

First, when reading an article or watching a video, check the source of the information. Determine whether the person or organization may have a bias. For instance, is it a credible news organization? Or is it a marketing campaign by a company that stands to profit from spin? Is it an unreliable blogger (not all bloggers are unreliable) who banks on the ad revenues of a huge following and thus has the incentive to generate outrageous information? Or are the claims from substantiated sources?

Second, check out the other side of the argument. When we stumble across something that is outrageous, it is best to do further research and investigation, which may prove that the argument was right or that the situation is more complicated. It may even yield an idea on how to improve the situation. There is nothing wrong with sitting on the fence and getting the bigger picture before choosing a side. In fact, it is encouraged.

Third, don't become overinvested in social media as a reliable source of information (a reliable source of entertainment, perhaps). As mentioned earlier, most popular social media outlets have become an echo chamber of our own thoughts and opinions. Many have evolved from what they once were when the Internet was yet new to now become profitable businesses that control content exposure through algorithms and show us tailored advertisements based on the digital imprint they have collected on our interests and former Internet activity. When we see things that paint groups of people with broadly negative characteristics it should arouse our suspicion toward the messenger as well as the message being delivered.

Fourth, enjoy conversations with family, friends, and colleagues, but don't always trust anecdotal evidence. When it comes down to it, we get most of our information from the conversations we hold with people we know. Because we know them, it is far easier to trust what they have to say. Oftentimes, what they share is accurate, but sometimes their stories and advice may be a bit off the mark or simply a one-off experience unique to them. Our family, friends, and colleagues are as imperfect as we are, and one conversation with one individual shouldn't shape an entire opinion.

Fifth, just because it glitters doesn't make it gold. We may adore our actors and musicians. We may be endlessly entertained by our YouTube divas and favorite bloggers. Yet just because someone is famous does not mean they are an expert. They can get it wrong just like we can, or they may have financial arrangements to support the interests of a sponsor. If being famous is their area of expertise, chances are food security isn't. Take what they say with a grain of salt.

Sixth and lastly, beware of anything that openly attempts to manipulate the emotions. This can include videos with heart-wrenching music or images that make us cringe, or stories of horrific despair that make us want to cry, or tales of extraordinary success that make us want to do or pay whatever it takes to obtain that success for ourselves. Not everything that prompts emotional responses is dubious, but many marketing campaigns and communication efforts purposefully play on our emotions so we will skim over the details. They know how to tap very effectively into our powerful emotions of fear and outrage, which will fog our rationality and motivate us into taking action (such as buying their product or giving them money).

Making an effort toward improving our communication through these steps places us on a good path. If the six points are too hard to remember, the following neatly summarizes the process: take a deep breath, slow down, and check the sources.

As a society, we can benefit from working toward broader communication efforts above and beyond the personal. There are two primary ways in which we can do this. First, we can purposefully build alliances with those we perceive to be different from us through our organizations and associations and local governments. Progress requires that we communicate with and engage with those beyond our immediate sphere of comfort. Food and agricultural groups cannot only engage with other food and agricultural groups. Scientists cannot only engage with scientists. Interest groups cannot only engage with like-minded interest groups. These efforts to work across differences must be genuine, however. They can't be efforts to bait the other side into a media gaffe or be used as an opportunity to confront people who don't share our views. Rather, they must be born from a desire or an effort to see an issue clearly through the viewpoint of another person.

Second, we can make a concerted effort to place sound, objective educational and actionable knowledge on food and agricultural issues into our classrooms. From elementary school through high school, in most schools the food and agricultural curriculum can stand to be built up and improved to become more comprehensive and encompassing. This may require support and effort from the broader community, because many schools are underfunded and don't have the necessary resources to achieve this on their own.

Looking Ahead

Poor communication around food and agriculture yields dire consequences—we fail to make the best decisions for ourselves, our communities, and the world. Ultimately, the hardest hit are the poorest of the poor, the ones who struggle to obtain food security on a regular basis and other marginalized groups who are not able to make their voices heard in our debates around food and agriculture.

In 2008, voters in California passed a ballot initiative called Proposition 2. It required that animals be provided room to turn around, lie down, stand up, and fully extend their limbs. Summarized in one sentence, this sounds great, but it was not without controversy. In the end, more than $15 million[11] was spent advocating for and against this proposition. In November of that year, it passed.

Egg farmers and California consumers had seven years to prepare themselves for the change (other industries were not as heavily affected because the chicken industry is one of the state's leading animal industries). When 2015 rolled around and the bill had to be implemented, the additional costs of changing the cages prompted the price of eggs in California to temporarily rise, also influenced by a bout of avian influenza in the Midwest.[12] Hardest hit by these price increase are the poorest consumers. For those already on a tight budget, eggs are traditionally a cheap source of protein. The spike in egg prices created the largest budgetary strain for California's lowest wage earners.

Perhaps this temporary outcome, and others similar to this, could be ameliorated through improved, dedicated communication among all parties involved and a willingness to find compromise. Compromises can sometimes provide gains for both sides, in this case better conditions for animals and food security for everyone, yet they are challenging because they require an elaborate process of communication for groups to establish and build on core shared values.

Communication is the hinge upon which decisions, both good and bad, rely. It can yield tremendous progress, and, if we make the effort to both speak and listen carefully about the challenges outlined in this book, we will be well on our way to solving them. However, if we don't invest wisely in our communication, the way we talk about food won't just fail us, it will also fail those who have no voice at all. And these people, as the next chapter illustrates, face dire challenges indeed.

Chapter 12

Achieving Equal Access

Gerald Shively

Ensuring that everyone, everywhere, has enough to eat

Today, the 4,000-foot runway in Tumlingtar, in eastern Nepal, is paved. But when I arrived there from Kathmandu in 1985 aboard a Twin Engine Otter, the runway was nothing more than a grass strip. With only 20 passengers onboard the tiny plane it didn't take us long to disembark and collect our luggage. Among the cargo was what we, in the developed world, might consider unusual: a 100-pound sack of rice. My friend Nemat was planning to haul it up the Arun Valley to his home village of Kuwapani. Rice, although a basic staple throughout much of Nepal and the world, was hard to get in Kuwapani, high in the Himalayas.

Nemat's desire to bring a sack of rice for his family was both a symbolic gesture and a food security strategy. Rice remains a relative luxury in remote mountain districts of Nepal, even 30 years later, because farmers in such locations can't grow rice, at least not very successfully. The growing season is too short. Rainfall is unreliable.

Fertilizer and improved seeds are costly, if they are available at all. Now, as then, farmers in Nemat's village mostly grow corn, potatoes, and wheat, while raising a few chickens. A well-off family might have a cow or a few goats. Diets are simple, and for the most part reflect what can be grown locally.

Although food shortages are relatively rare in and around Kuwapani, especially compared to other parts of Nepal, local diets lack diversity. This tends to create problems from a nutritional point of view. We as humans need a wide range of nutrients to thrive, and we usually gain these through eating a varied diet. How many of us in the developed world eat the same meals day in and day out? By choice, relatively few.

Yet many of the residents in Kuwapani, especially young children and women of childbearing age, eat a monotonous diet and fail to obtain sufficient protein and micronutrients for robust health and physical growth. This is a problem across much of Nepal, where rates of childhood malnutrition are among the highest in the world. In 2011, the most recent year for which national data are available, between a third and half of *all* Nepalese children had stunted growth, a largely irreversible condition reflecting the cumulative effects of chronic undernutrition and poor health.

The importance of proper early nutrition cannot be overstated, and really gets to the crux of the matter regarding food security. Everyone, especially young children, needs adequate and proper nutrition. Food is often within reach, but access remains unequal, with the wealthy enjoying abundance, and the poor suffering from scarcity. Most of us can agree that starvation and malnutrition are abhorrent, yet they persist. How can this be?

Before addressing that question, let's consider the focus on children. Although scientists continue to study how malnutrition and child development are connected, three things appear clear.[1] First, much of the brain's capacity and structure are already determined by age three, and the brain's development is correlated with a child's outward physical growth. This is a primary reason why stunting is used to measure malnutrition. Using a reference growth curve, nutritionists identify children as stunted if their height falls below that of 95 percent of well-nourished children of the same age. Those kids are at high risk.

This risk is also why interventions for reducing malnutrition are often targeted at children during the first thousand days following conception. Brain systems regulating mood begin developing even before birth, those in charge of attention and multitasking have a growth spurt in the first 6 months following birth, and the regions governing spatial understanding grow most rapidly in the first 18 months after birth. During these and other critical periods, nutritional deficiencies can seriously undermine development.

Second, not all foods are created equal. The micronutrients iron, zinc, and iodine are all essential for brain development, and many low-income diets heavily dependent on coarse grains and starchy roots are deficient in all three. These micronutrients are so important that some researchers have suggested that ridding the planet of just these three deficiencies alone would raise the world's IQ by 10 points.[2]

Third, although young brains are developmentally flexible, and mental stimulation and interaction with caregivers can sometimes compensate for lack of good nutrition in a growing child, catch-up is extremely difficult. The possibility of recovering from nutritional deficits generally diminishes over time. Studies have shown, for example, that growth rates before 12 months of age accurately predict cognitive performance at age nine, and some research suggests that early nutritional deficits may be correlated with reduced cognitive capacity in later adulthood. What does this mean? Simply put, wherever poverty drives early childhood malnutrition, it also tends to perpetuate itself across generations. Poorly nourished children grow up to become less successful than they otherwise could, increasing the likelihood of passing on their own poverty and underachievement to their own children.

Intervening during early childhood can be an effective strategy to break this vicious cycle. Yet malnutrition and food insecurity, like the other issues in this book, have complex causes. The "how" and the "why," which we explore in the remainder of this chapter, are difficult questions, to say the least. What's more, overcoming malnutrition and food insecurity often depends on meeting all the previous challenges explored in this book. In many respects, achieving equal access to food sits at the intersection of many other challenges. And by allowing food insecurity and unequal access to nutritious food to persist, we leave a tremendous amount of human capacity unrealized. Tapping that unrealized potential could help solve many of the world's other vexing problems.

Unable to Grow, Unable to Purchase

Malnutrition and food insecurity begin at the household level, when people are unable to grow nutritious food, unable to buy nutritious food, or both. If someone in Kuwapani is eating rice, chances are it was grown elsewhere, bought with income earned elsewhere, and transported there on roads and bridges built by the government or with foreign assistance.

In that sense, a simple sack of rice embodies the key components that must come together to ensure food security at any given place and time. People need access to nutritious food, and they must be able to afford it when they need it. Luckily for his extended family, Nemat was earning a reasonable wage working in the Kathmandu valley. He could afford to purchase some rice and hire a porter to transport it back to his village. Even though rice is not a nutritional powerhouse, Nemat's sack added a few months to the family's supply of basic carbohydrates. This was probably enough to get them through the lean season before the local harvest, and to provide some modest insurance against local crop failure, putting them in a much better position than some of their neighbors.

But even an adequate supply of rice, which provides plenty of carbohydrates, is not enough to forestall malnutrition. No matter how much rice or corn you eat, if all you eat is rice and corn, you won't be adequately nourished. A common myth is that malnutrition reflects lack of food. But it is not enough to have food, it must be the *right kind* of food.

A second popular misunderstanding about hunger and malnutrition is that food insecurity arises from widespread food shortages. In the modern era, at least, that is simply incorrect. As the Nobel Prize–winning economist Amartya Sen pointed out in his study of the 1943 and 1974 Bengal and Bangladesh famines, food is often available, even where people are starving.[3] What *acutely hungry* people lack is either the land to grow food or the income to buy what they need. This is what blocks them. When food is there, but they haven't grown it, they starve if they don't have the money to buy it.

When people face this situation over a long period and cannot obtain a consistent supply of essential nutrients, they become *chronically malnourished*. Chronic malnutrition is what ultimately leads to stunted growth in childhood. The body diverts available nutrients away from linear growth and toward basic survival.

Tragically, as many as one in four children worldwide are stunted, and roughly half of all deaths in children below age five can be attributed to malnutrition in some form. Such hunger and malnutrition currently exist even though the world produces enough calories to provide sufficient energy for everyone on the planet—roughly 2,900 calories per person per day according to the latest estimate from the Food and Agriculture Organization (FAO) of the United Nations.[4] Whether future supply will meet future demand is a concern, of course, especially as incomes grow and consumers seek to add more meat and dairy products to their diet. These foods require large amounts of grain for feed, which in turn places pressure on land and water. Still, at least for the time being, the world's overall supply of food seems sufficient to meet our basic needs.

So why do people continue to go hungry? In far too many settings, civil unrest undermines food production and distribution. Although famines are relatively rare, they continue to occur, and a large proportion of the famines that have occurred over the past 25 years, as well as current food crises, can be traced to war or armed conflict. Perhaps the single most effective strategy to end acute hunger worldwide would be to establish peace in war-torn areas.

But in a peaceful place like Kuwapani, food insecurity has more subtle causes. Hunger and malnutrition in such locations reflect a lack of production diversity, seasonal shortages of things that are locally produced but not easily stored, high prices for food items that are not locally produced, and poverty due to limited economic opportunity. These same conditions also point to potential solutions: home gardens that diversify production; better management of food waste and loss; greater access to food markets; cheaper and more reliable transportation; basic nutrition education; and higher household incomes.

To repeat, the problem confronting the majority of hungry or malnourished people is either insufficient access to nutritious food or insufficient purchasing power to obtain that food, or both. Over time, basic economic development can help alleviate the situation, but economic development takes decades. In the meantime, multiple generations of malnourished kids miss out on developmental gains, growing up to achieve less than their potential, and passing on poverty to their own children. The challenge is to find targeted ways to accelerate progress on a broad scale, and not simply wait patiently for the general process of economic development to improve nutrition and health.

Although it is easy to imagine the situation in Kuwapani as purely a third-world problem, the truth is that similar patterns arise in urban areas and extremely rural areas of affluent countries. In many places, household budgets are tight and options for purchasing healthy and fresh foods are limited, resulting in pockets of concern that have come to be known as food deserts.[5] In the United States, you can find these food deserts in many large cities, among them Chicago, New York, and San Francisco. However, the nutritional consequences of living in an urban food desert differ from those found in rural areas. For poor urban residents of high-income countries, the common health problems are those that arise from consumption of highly processed and nutritionally weak foods. The ills include obesity, cardiovascular disease, and diabetes. Despite the differences in consequences, however, potential solutions to the problems associated with urban food deserts sound surprisingly similar to those that one might prescribe for Nepal: community vegetable gardens; better access to food markets; cheaper transportation; basic nutrition education; and higher incomes.

The Importance of Prices and Incomes

The *New York Times* headline on February 21, 1917, read, "House and Senate Debate Food Riots: Spectacle of Hunger in Richest Country's Richest City Stirs Demand for Inquiry." The debates in question, which took place in both the House and the Senate, focused on alleged manipulation of food prices and widespread unrest in response to a rapid run-up in the prices of basic staples. Similar protests against high food prices occurred the following year in Japan, where rioters used dynamite in their attacks against the homes of the rich.

It is hard to imagine now, but a century ago, US and Japanese citizens, especially middle and lower income Americans and Japanese, spent a large fraction of their income on food. As a result, even modest increases in food prices had strong political and social repercussions. In contrast, when food prices rose sharply around the world in 2007 and 2008, there was plenty of grumbling in America and Japan, and true hardship for some groups in these and other rich countries, but little widespread unrest. Instead, the major food riots of 2007 and 2008 occurred in Africa, the Middle East, and South Asia. In Egypt, the world's largest consumer of bread, the price of a loaf increased fivefold in early 2008, leading to lines outside bakeries, tense scuffles, and sometimes-violent clashes that helped fuel the popular uprising known as the Arab Spring.

We can better understand public reactions to price increases, and the private impacts they exert on pocketbooks and dinner plates, by considering what economists call budget shares. A budget share is simply the proportion of a dollar of income that a family spends on some category of purchases, say housing, clothing, or food. The following graph displays the budget shares for food (including beverages and tobacco) for 144 countries in 2005 (fig. 12.1). I constructed the chart using data from the US Department of Agriculture (USDA) International Comparison Program, an undertaking that attempts to compile and analyze in a consistent way household income and expenditure data from across the globe.[6] As calculated by the USDA, food budget shares range from a low of 5.7 percent in the United States to a staggering 63.4 percent in the Democratic Republic of Congo (DRC). In the DRC, on average, a family spends 63 cents out of each extra dollar of income on food. When prices rise in the DRC, people really feel it. Not surprisingly, a century ago, when US consumers were rioting, food budget shares in the United States were much higher than they are today. In fact, at 43 percent they were roughly on par with Nepal today and substantially higher than those in Egypt. As a result, food prices were much more important to American consumers in 1917 than they are to average Americans today.

The Nutrition Transition

An obvious pattern underlies figure 12.1. In the chart, each dot represents a country. As points of reference, I've labeled a few and ranked the countries from rich (at the top) to poor (at the bottom). The data clearly illustrate how higher incomes are associated with smaller food budget shares. The rich spend a smaller share of their income on food than do the poor, a pattern that holds whether one compares rich and poor consumers within a country or rich and poor countries across the world. The reason is that there is a physical limit to how much food a person can eat, and therefore an economic limit on how much someone can spend on food. Even if one of the world's 1,800 billionaires decides to dine on Almas caviar ($10,000 per pound at last report), there is an obvious upper limit on the amount of caviar someone can eat. Eventually, she (there are 167 women on the list) has to spend her money on something else. And even for nonbillionaires, as incomes rise, things other than food begin to look appetizing.

Although a broad range of factors influences food choices, among them culture, taste, convenience, and advertising, numerous

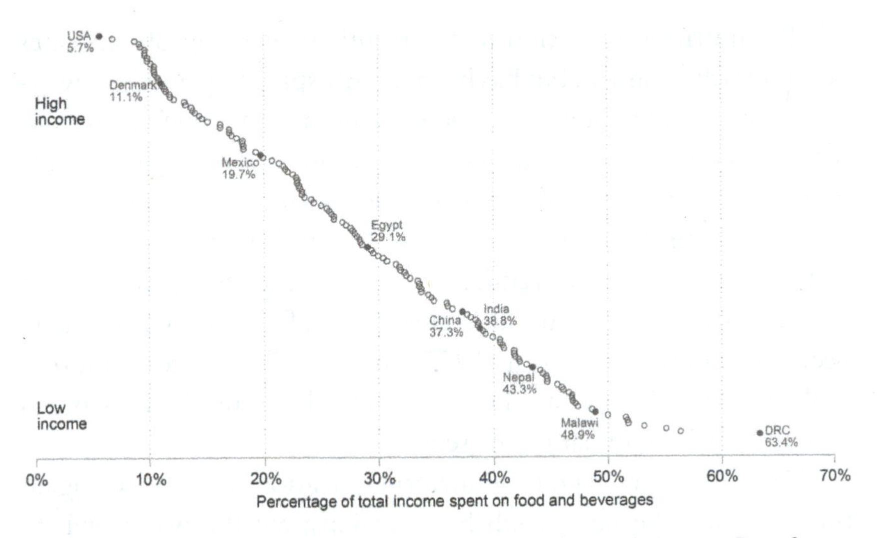

Figure 12.1. Food budget shares for 144 countries in 2005. Source: Data from the US Department of Agriculture International Comparison Program.

studies have shown that changes in food prices and incomes have important implications for the kinds of food choices people make and, by extension, their health and nutrition. Just as billionaires seem to develop a liking for caviar, the rest of us begin to shift our consumption away from unprocessed grains toward animal products and processed foods as our incomes rise. Researchers refer to this shift as the nutrition transition, a term coined by the American scientist Barry Popkin.[7]

Simply put, the "nutrition transition" refers to the way people gradually shift away from inexpensive diets low in calories and nutrients toward diets that are higher in calories, and then toward more costly diets that are more nutritionally balanced and contain more protein from animal sources. It might seem as if the evolution in the structure of diets would stop there, but the transition continues to unfold, as people consume more and more meat and dairy, more and more processed foods, more and more meals away from home, and more and more junk food and fast food. The combination of changing diets and changing lifestyles leads to rising rates of obesity and associated health risks.

The nutrition transition that currently fuels rising obesity rates began in rich countries but has been rapidly spreading and is now evident even in poor countries. Where it will end is anyone's guess, but some observers have speculated that the transition may eventually move in a more favorable direction, as consumers in high-income countries begin to recognize the negative health effects of modern diets and start to limit or reduce consumption of meat products and highly processed foods. In the United States, for example, per capita beef consumption peaked in 1977, and overall meat consumption peaked in 2006. Both have been gradually declining. By how much and for how long remain to be seen.

To better understand this nutrition transition, we can once again turn to data. The next graph helps to illustrate the way in which the overall quantity and specific composition of diets change over time, as countries and their populations become more prosperous (fig. 12.2). I constructed this graph using "food balance" data covering the period from 1961 to 2013.[8] These data come from FAO, which first began compiling country-level data on food supplies following World War II in response to acute food shortages in Europe and elsewhere. To make comparisons across places and time, I used the food balance data to summarize the historical situation in the United States and five other geographic regions: the European Union, South America, South Asia, East Asia, and Africa.

At the most basic level, food balance data measure how much food is available to people living in a particular country at a particular point in time. To create the data, FAO statisticians start with their best guess as to the total quantity of food produced in a country in a given year. They then add to this the total quantity of food imported from other countries, and subtract the amount exported. They make additional adjustments for quantities held in storage, or amounts used for other purposes, such as animal feed. This gives a reasonable estimate of the amount of food available in a country. More production or more imports mean more food. Less production or more exports mean less food. You get the picture.

The FAO then divides this total amount by the country's

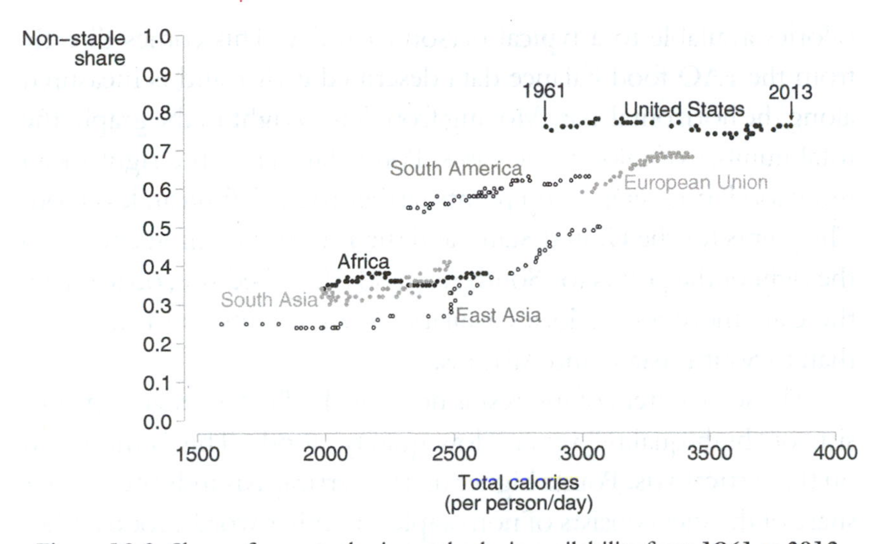

Figure 12.2. Share of non-staples in total calorie availability from 1961 to 2013. Source: Data from the Food and Agriculture Organization of the United Nations.

population to arrive at an annual estimate of per capita availability. The result is a rough but reliable snapshot showing the amount of food *available* in a country.

Because the FAO collects data on a very wide range of individual food items, we can use the information to compare diet composition across countries and over time. The data are not perfect, however, because they reflect national aggregates of food availability and say nothing about how that total amount is distributed across the population. The data are also silent on the exact amounts *consumed* by specific individuals or groups of individuals, which is what really matters for nutrition and health. Nevertheless, the data turn out to be very useful for understanding the food situation in a particular place. They also provide a strong foundation for discussion, policy analysis, and government decision making related to agriculture and nutrition.

This graph showing total calories and non-staple shares can tell us a number of interesting things about typical diets and also helps us to understand what has been happening over time. The first piece of useful information the graph provides is the average number of calories available to a typical person each day. This comes directly from the FAO food balance data described earlier and is measured along the horizontal axis. Moving from left to right in the graph, the total number of calories increases. Points farther to the right mean more food, on average, and points farther to the left mean less food. The points for the United States and the European Union end up to the right of the points for South Asia and Africa because, on average, there are more total calories available to Americans and Europeans than to South Asians and Africans.

The second item of interest is how much of the average diet consists of "high-quality" versus "low-quality" foods. This is measured on the vertical axis. Points higher up the vertical axis indicate a larger share of the diet consists of non-staples, in other words, foods other than coarse grains and starchy roots (think unprocessed rice, corn, potatoes, and cassava). To obtain this information, we simply add up individual food items that appear in the FAO food balance data. Moving from the bottom of the graph to the top, the quality of the average diet improves, in the sense that a larger proportion of the calories available comes from more nutritionally dense food sources. The points for the United States and the European Union lie above those for South Asia, East Asia, and Africa because, on average, a smaller share of the calories consumed by Americans and Europeans comes from basic staples.

The third and final piece of information contained in the graph is an indication of how diets have changed over time. To see this, first recognize that each dot in the graph represents the food situation in a particular country or region in a given year. Because total calorie availability and total income tend to go hand-in-hand, the graph shows how food availability and diet composition have changed as countries have become more prosperous over time. For each series of points representing a region, a movement from left to right is a movement over time from 1961 to 2013. Data points for the United States appear in the upper right area of the graph. This is because American diets contain both a large number of calories and a large share of non-staples.

The data points for the United States show that the share of non-staples in total calorie availability has been relatively flat over the past half century. The United States had already moved through the early stages of the nutrition transition by 1961, with the share of staples in the average American diet settling at roughly 20 to 25 percent of total calories. The European Union transitioned a bit later and seems to have stabilized at a somewhat lower share of non-staple calorie consumption. In contrast, South Asia and Africa exhibit low total calorie availability over the period, and heavy reliance on basic staples. The evolution of diets in East Asia has been the most dramatic. There, total calorie availability doubled from 1,500 calories per person per day in 1961 to more than 3,000 calories per person per day by 2013. In terms of calories and composition, East Asian diets now look quite similar to European diets from 40 to 50 years ago. That's the good news. The bad news is that obesity rates in East Asia are catching up too—and fast.

What factors explain shifts in diets over time, and in particular the remarkable shift in diets in East Asia over this period? Rising incomes drove the trend. But behind the growth in incomes were other factors, including rapid innovations in agriculture that led to substantially higher crop yields. These innovations were fueled by public investments in agricultural research as well as the rapid spread of technological advances that proved useful. Urbanization, the expansion of market infrastructure and transportation, the growth of regional and international trade, and the growth of nonfarm sectors of the economy also played important roles. These are the same forces that will be needed to lift calorie availability and improve diet quality in Africa and South Asia, in places like Kuwapani.

The Next Wave

If you find yourself in Copenhagen, with an evening to spare, a friendly Dane might point you in the direction of Nørrebro, a hip neighborhood on the near-west side of the city, not far from the public cemetery where the storyteller Hans Christian Andersen, the philosopher Søren Kierkegaard, and the American jazz saxophonist Ben Webster are all buried. One of the main avenues, Jægersborggade, was once a rough place that the well-heeled largely avoided. But in recent years it has gradually gentrified, with art galleries and new restaurants popping up along the street at regular intervals. Anchoring the neighborhood, at number 41, is the Michelin-starred restaurant Relæ, which for two years running has been named the world's most sustainable restaurant.

In addition to providing a menu of cutting-edge Nordic cuisine, which you might expect, Christian Puglisi, Relæ's owner, produces something you might not expect: an annual sustainability report.[9] In it, he describes in detail the restaurant's sourcing of ingredients, their choice of light bulbs and furniture, the cleaning products they use, the energy efficiency of their stoves, and the fact that they use a bicycle to pick up their bread each day. Along with a discussion of local ingredients and supply chain efficiencies are comments on healthy eating and responsible marketing. Relæ's most recent innovation is to revive an old farm on the outskirts of Copenhagen, where they are growing their own organic vegetables and establishing a micro dairy to supply the restaurant's needs. All of these are private actions aimed at ensuring quality food and enhancing the sustainability of the food system.

Thinking back to our latest graph, we might reflect a bit on patterns and trends, and speculate about the potential future of food and global food systems. If present trends continue and incomes continue to rise, consumers in Asia, Africa, and Latin America will begin to more closely resemble average consumers in places like Copenhagen. They will continue along a path of greater calorie consumption and larger non-staple shares. To keep up, the world will need to produce more calories, many of them in the form of animal protein. In coming decades this nutrition transition will place a strain on agricultural systems and the natural systems on which they depend. We are already starting to see this happening in many places and are beginning to recognize that our demands on soils and water may exceed nature's capacity to provide. At the same time, dietary changes are creating public health challenges, as the prevalence of noncommunicable diseases rises in step with rates of obesity. In many respects, the unfolding nutrition transition looks a lot more like a slow-motion nutrition train wreck. Can we apply the brakes?

One response, of course, could be a shift in Western diets toward food that is more healthful and sustainably produced. That is part of what's going on at Relæ and restaurants like it, and it is not inconceivable that Western diets and Western food systems will evolve in healthier and more sustainable directions. Things like community supported agriculture and farm-to-table restaurants may become commonplace not only in countries that are currently rich but also in countries that are soon-to-be-rich. And while it seems extremely unlikely that one trendy restaurant on an alleyway in Copenhagen can change everything, part of the answer to sustainably feeding a growing planet is likely to be found in a reinvention of diets, and a reorientation of food systems. This, in short, was the conclusion reached by the World Health Organization (WHO), which in 2014 looked carefully at the European population's diet and nutritional status with an eye toward reducing preventable, diet-related, noncommunicable diseases.[10] The WHO argued in favor of policy options for rich-country governments that focus on influencing "the production, marketing, and availability and affordability of foods."

And yet while that might seem to be a viable approach for action in the rich world, what about the poor world? Where people are still poor, something more is probably needed.

Twinkle-Twinkle Little Star

Every 99 minutes, day-in, day-out, a NASA satellite crosses the equator from south to north as it circles 438 miles above Earth's surface. The Aqua satellite silently orbits our planet, staring down dispassionately on our growing population. At the moment it was launched, on May 4, 2002, Earth's population stood at roughly 6.2 billion. Since then, we have added another billion. Every second, on average, we add two new people to the human family. With every pass of Aqua, that's 10,000 more mouths to feed.

Aqua is a real workhorse.[11] There are six instruments on board the satellite, one of which is a curious digital camera called the Moderate Resolution Imaging Spectroradiometer (MODIS). MODIS can see 36 frequencies of light, and every day it snaps pictures of Earth's surface with a field of view 1,400 miles wide. Given its orbit, this means MODIS sees nearly every inch of the planet every day. Among the things that MODIS notices is how green places are looking down here, a characteristic that scientists quantify as the Normalized Difference Vegetation Index (NDVI). As locations subtly shift from brown to green to greener and back again, month after month and year after year, the NDVI information begins to give scientists a sense of what looks normal across the seasons in an area, and what looks less than normal. In fact, MODIS can often see trouble ahead—say, a drought—before people on the ground are even aware of what is happening.

One problem that is hard to ignore, and one that MODIS can see and is helping to solve, is that kids in Kuwapani, and millions of kids like them elsewhere in Nepal and other parts of South Asia and Africa, are highly sensitive to local growing conditions. Many of the kids who live in rural areas of the world are the children of farmers, and what their families can grow largely determines what they can eat. The conditions for agriculture, therefore, have an important influence on the conditions for child growth. We know this, because when we line up the satellite data with the data on child growth, we see evidence that children's weights and heights are correlated with the satellite signal: more green generally means healthier kids.[12] This underscores the importance of continued efforts to improve agriculture in less favored parts of the globe. Where crops are healthier and farms are more productive, people are too. Evidence also suggests that investments in infrastructure are important. More roads and bridges connect people to markets. More hospitals and clinics improve maternal and child health. Both kinds of infrastructure can help to buffer children from the uncertainties of weather and from agricultural shortfalls.[13]

One way that MODIS is helping to solve food security problems is by increasing the amount, accuracy, and timeliness of data on agricultural conditions. This helps in multiple ways. First, it can help to improve the targeting of interventions when things go wrong. As an example, MODIS provides data for the Famine Early Warning System Network (FEWS NET).[14] FEWS NET is designed to improve the way rich countries send assistance and aid to places where it is needed, and information from MODIS can help to reduce delays and inaccuracies in targeting. This saves time, money, effort, and lives. The kind of data provided by MODIS can also help to improve how the public and private sectors invest in agricultural interventions and technologies, and can support better outreach and training assistance to farmers by targeting efforts where and when they are needed. Although MODIS is more than a decade old, it is still generating useful and valuable data. That is pretty incredible. But what is even more incredible is the range of new technologies that are becoming available to help us understand our physical environment. Using satellites, drones, and cell phones, new technologies are providing useful real-time information on everything from the moisture in soils to the density of insects in a farmer's field.

If you pause for a minute to think about how rapidly technology is evolving, it is impossible not to be optimistic about the potential for a twenty-first-century technological revolution in agriculture, provided we invest in it. The main challenge going forward will be to recognize that all human beings need and deserve proper food and nutrition to grow and flourish, that it is in everyone's long-run best interest to improve global food security, and that these new technologies need to reach every corner of the globe to do so. In the coming decades, we will undoubtedly have the technologies and the knowledge to meet the challenge of nourishing the world. To make progress, we will need to continue doing what works, namely delivering the basic things that families in places like Kuwapani need:

diverse, productive, and reliable crops; greater access to food markets; reliable transportation and communication; basic education; and greater economic opportunity. But to truly succeed, we will also need to ensure that people everywhere, including people like Nemat and his neighbors, have access to innovative technologies that can unlock hidden potential.

Ensuring that people everywhere are properly nourished is a challenge that ties together everything we have explored in this book. It links climate, technology, food loss, water, population, and more. Yet despite the complexities of this challenge, we *do* have the capacity to nourish everyone, and the implicit motivation to do so. We just need to make meeting the challenge a priority. In the end, that comes down to us. The matter lies, quite squarely, in our hands. The question is, will we rise to the challenge?

Conclusion

Jessica Eise and Ken Foster

I, your editor Jessica, sat in on a lecture by Brigitte Waldorf on population about a year before we wrapped up this book. Brigitte went over a number of grim scenarios. The class, myself included, all felt a distinct sense of alarm. Yet Brigitte herself was downright blasé. Her lecture could even have been described as *cheerful*.

Toward the end of the class, while looking at what appeared to us to be insurmountable projections for population growth before it would ultimately plateau, she seemed to get a sense for our concern. With a smile, she told us, "Don't worry. I'm optimistic."

Optimistic? How could she be optimistic? We looked at one another with confused and disbelieving faces. A brave student chimed in, "But how? How can we manage such population growth?"

I'll never forget her response, and do my best here to faithfully recreate it. "I have faith in humankind. We have never failed to meet a challenge yet, and I trust that we will do the same once more. People have been making dire predictions for centuries, and each time we find a way to overcome. I have confidence in our ingenuity, and our ability to find a way," she said, with an implacable and comforting confidence. "Plus, I'll be dead by then, anyway," she quipped, a glint of humor in her eye. We laughed.

Right now, this very day, the challenges we face extend well beyond Brigitte's population projections. As we've seen on the earlier pages of this book, each challenge is complex in its own right. What's more, these challenges are also interconnected. This interconnectivity is an important component of solving them.

Population was, however, the natural starting point for the book. It is the underpinning of so many of the challenges we face. With more and more people, there are more and more pressures weighing upon us in our quest to feed the world well and sustainably. Yet, as our author Brigitte Waldorf explained (the very one and the same who gave that fateful lecture a year before publication), the population equation is not so simple as it appears. And when it comes to feeding the population, the intricacies skyrocket. Where exactly does the person live? Is said individual mobile? Is he or she rich or poor? Young or old? Disenfranchised? Empowered?

Yet population just sets the stage. Our growing numbers not only challenge the quantity of food we produce, they also spill over into environmental impact, particularly concerning the management of key water resources and the ways we use land, many areas that were largely taken for granted by past generations. We no longer have this luxury. The next two chapters (2 and 3) following the population discussion focused on our precious commodities of water and land. Whole books could be written on the interaction between humankind and the environment. Our authors adeptly focus on the agricultural and food issues related to these essential resources.

We have only to glance, for instance, at a globe to see that blue water stretches across and around our Earth. Yet with all this water covering our planet, only a small fraction of it is useable for humans and that fraction is distributed unevenly across the globe. Yet demands for water and overextraction of groundwater, with little thought to the future, are occurring across the United States and the world. Many are waking up to the reality that water is a precious and limited resource, yet even if one is lucky enough to live where it is abundant, this is no consolation if its quality is ruined beyond safe use.

How we manage our water resources in the present has and will have irreversible consequences for our food security and well-being within anyone's lifetime. Laura Bowling and Keith Cherkauer showed us this in chapter 2, as they took us on a global journey of water, from conflict zones, regions with water scarcity, the water footprints of food, all the way through the rivers, lakes, aquifers, and other water sources. We face a dangerous future if we fail to use this precious and oft-overlooked commodity wisely. Water is essential for growing crops on our precious land and soil resources. Anyone who has forgotten to water their house plants, or lived through an extended drought, knows this firsthand.

The complementary nature of water and land in food production guided us to our next chapter, chapter 3, which focused on land. Land is a finite resource, and arable land is even more precious still. There is no more land to be mined, built, or bought once we've occupied it all. The geological processes that grind rock into clay, silt, and sand then mix them with thousands of years of organic plant matter to create soil are measured in eons. Once land is degraded through human activities via erosion, pollution, salinity, or development, it is extremely difficult if not impossible to reclaim it for farming and to do so sustainably.

Today, there is increasing competition for farmland. Growing populations lead to urban sprawl and denser highway systems that sometimes take land permanently out of other uses, including food production. In chapter 3, Otto Doering and Ann Sorensen explored the historical evolution of land use and the increasing threats we face today, as well as where we can, or might, go in the future. As we speak, there are forces at work changing land use patterns. Some are at a local scale, others on a global one. The overwhelming global culprit is, however, climate change. Farmers must change what crops they grow in which parts of their local counties. But more importantly, the band of latitude in which the world's major food crops—grains and beans—grow is marching northward and southward while the patterns of rainfall in the tropical and subtropical zones are becoming less predictable, placing some of our favorite foods like bananas and chocolate at risk of drought, disease, and pests.

It should come as no surprise, then, that our fourth challenge was the interplay of a changing climate and agriculture. As Jeff Dukes and Tom Hertel carefully unfolded in chapter 4, climate change will significantly impact our ability to feed the world. If temperatures are too hot, plants and animals as well as the workers tending to them suffer greatly. If there is too little rain, or too much, crops will fail. Even if rainfall patterns are not affected but temperatures rise significantly, water is lost before crops can take advantage of it.

The consequences of climate change are manifest every year as the world breaks its previous record for global temperature. The trend of rising heat-trapping, greenhouse gas emissions is not expected to reverse any time soon. Today, farmers in the Midwest are planting earlier but finding themselves subject to more large rain events that lead to replanting, delayed planting, or drowned crops. Can farmers continue to adapt to the new climate and its increasing vagaries? Will temperatures continue to rise indefinitely? Is there something that can be done to slow the changes? How will policies affect farming, which is, in itself, a significant source of greenhouse gases?

Nature has been silently enduring the unintended consequences of certain technologies. But many of these technological developments have helped us to obtain one of the greatest achievements of humanity in the past half century, that of feeding today's seven billion people. This meteoric rise in agriculture also made food more affordable and accessible, lifting millions out of hunger. Technological innovation, the topic of chapter 5, has given us the means to obtain monumental accomplishments.

Yet, to face the challenge of feeding the world in the era of climate change and greater land and water limitations, we need to rethink how we produce our food. Technology is what enabled us to feed today's population, and it remains the key to developing more sustainable farming innovations and smarter food production methods.

However, there are signs that agricultural productivity growth is slowing, and it takes time for new ideas to expand from the laboratory to the field. In chapter 5, Uris Baldos covered the reality of technological innovation and food production from its inception to today, concluding with the emerging technologies that can help us attain our goal of sustainably feeding the world through a "greener" green revolution, if we act now.

Technology itself was what prompted the development of a vast, complex system of food production that can feed so many people so affordably. And while the challenges of population, water, land, climate, and technology all seem pretty clear at surface value, other challenges are not so readily apparent. Chapter 6 tackled a subtler challenge, that of balancing our various food systems.

There are lots ways that food can land on our plates. It can be as simple as walking out to our own garden and pulling something off the vine, or as complex as the journey a fresh piece of salmon makes to a landlocked state. It is also not just how it is delivered that can differ, but the way in which it was grown. Is it organic? Local? Conventional? Or perhaps GMO? Our food system is complex and diverse, and in this chapter an equally and wonderfully diverse group of researchers, Michael Gunderson, Ariana Torres, Mike Boehlje, and Rhonda Phillips, teamed up to review the ways and forms in which our food arrives at our plate. Following the journey of a humble tomato, they broke down the complexity of our food value chain from what we eat, how it gets there, and where and how it was grown. These researchers looked at the challenge and importance of managing such diversity of systems, and why choice really does matter.

When technology and food systems shift and change, it is often dramatic. It can even overwhelm local demand for the food being produced, because farmers can learn to produce so much more than they had before. Paradoxically, in these cases farmers can be their own worst enemies, because prices may fall to levels that are economically unsustainable for them as a result of their successful adoption of new ideas. These problems have become so serious at times that the US government has *paid* American farmers to *not* plant crops on a portion of their acres. It is easy to glibly suggest that farmers grow something else, but if they are really good at growing a particular crop because it matches ideally with their soil, climate, equipment, and abilities, then the world may lose if they switch to other crops they aren't as good at producing. What is the solution to this dilemma? It's called trade. It has become an important aspect of sustainably and affordably feeding the world.

Chapter 7 directly tackles the often-overlooked challenge of managing international trade for food security. Unlike manufactured products or banking services, agriculture is tied to climate, water, and soils. Some places are well suited to farming, others are not, and each locale is better suited for some crops rather than others. We don't grow bananas here in Indiana for a very good reason called winter. The only way certain regions can enjoy fresh fruit and vegetables throughout the winter months is to import products from the tropics or the opposite hemisphere. In a mutually beneficial way, people living in the tropics benefit from trade through access to high-protein and high-energy crops that are more efficiently produced in temperate latitudes like the midwestern United States.

What's more, international trade is a buffer against unforeseen weather and economic events. The 2001/2002 drought in Australia virtually wiped out rice production in the Murray–Darling basin. Yet Australians didn't starve. They imported rice from elsewhere, along with adjusting their diet. Trade was a buffer protecting them against the extremes and unpredictability of climate. Similarly, during the deep global recession of 2007/2008, far fewer people faced food insecurity because global trade in food and agricultural products generally remained robust, meaning that all but the most isolated and poorest were able to obtain food. So why, then, is international trade policy so contentious, if it protects us against unexpected catastrophe and gives those of us with means what we want to eat when we want to eat it? Opening up to world markets shrinks less competitive sectors and expands those that are more competitive. Yet, as Tom Hertel explained in this chapter, we don't have a choice if we want to survive in a world of increased interactivity, resource scarcity, and climate instability.

Trade buoys prices so that farmers are more diligent in harvesting their crops, and it keeps perishable crops from piling up in storage on farms where they may spoil or be attacked by pests. Trade has great potential for reducing the amount of food that is wasted or lost. Yet even with this buffer of trade, we still experience far too much food waste and loss to consider our food system sustainable. This brings us to the next up in line, the topic of food waste and loss, covered in chapter 8 by one of your editors himself (yours truly, Ken Foster).

Food waste and food loss are two very different things. One is an unplanned decision made by a well-fed nation of people. The other is a devastating force wrought upon populations who usually can't afford to lose their crops. Food waste happens at the very end of the food supply chain. We buy more than we need, and it goes to waste in our grocery stores, restaurants, cupboards, and refrigerators. Food loss happens throughout the food chain. It is desperately fought by farmers around the world. Crops are lost to uncontrolled weeds, bad weather, insect blight, poor storage, lack of harvest technologies, forces of nature, poor infrastructure, or even unstable political or economic situations. The statistics in this chapter may have surprised you in terms of the magnitudes and potential to address food security issues, and the chapter carries direct messages about what we all can do about food waste in our homes and communities.

Speaking of what you can do, in this case perhaps for yourself, chapter 9 addresses the unforgettable challenge of health. Food *is* one of the primary contributors to our health. Too little? We fall ill. Too much? We also fall ill. Just the right amount of the wrong stuff? Same story. Nutrition is a critical component of the well-being of our population. It keeps us healthy so we can be productive as well as keep our health care costs low. Yet obesity is on the rise and so are the related health problems that come along with it. Many Americans are confused about how they should eat, and eating is not easy when dinner is supposed to satisfy our hunger, prevent metabolic syndrome, and save our community and possibly the world. Science says one thing, the health gurus another, and it becomes a quagmire of confusing, exhausting choices. In this chapter, Steven Wu evaluated this problem using an "economic approach" that is simple, clear, and concise. He answers a number of questions: What are the two or three most important drivers of obesity? What are the trade-offs of eating healthy? And what is the most efficient way to eat healthy? Ultimately, we do something if the benefits outweigh the costs. Choosing what and how to eat is a monumental task, and this chapter looks at the challenges of doing so well.

Health, a very personal topic, was followed by yet another. Chapter 10, like food systems, was a subtle challenge, one that might not even appear to be such at first glance. Throughout the book, various solutions to our challenges have been discussed and suggested, including everything from grand policy initiatives to farmer decisions down to our personal behavior. In order for these to happen on the scale that matters, they need to be "socially acceptable." In this chapter, Nicole J. Olynk Widmar took us on an exploration of our influence as consumers to revoke and grant social licenses to operate, where we saw that cultural norms and the changing roles of animal species have major consequences for agriculture and food production worldwide. Using our power to revoke and grant social licenses to operate or to behave in particular ways carries consequences. We can do great things with it. We can also do harm. For instance, failure to understand and appreciate differences in opinion and the nuances of a situation has the potential not just to challenge and disrupt trade relations but also to fuel social discontentment. Interrupted trade and social discontent are serious things indeed when it comes to food security. And thus our capacity as a society to grant social licenses to operate faces a unique challenge. Can we do so responsibly, informed and with an attitude of tolerance toward different preferences? Or will we become so stubborn in the righteousness of our beliefs that we disrupt, interrupt, and diminish our capacity to feed the world?

Behind these decisions to revoke or grant social licenses to operate lies the mighty force of communication, or the endless flow of information between us and the greater world. In chapter 11, your other editor (Jessica Eise) stepped up to cover our current state of communication around food and agriculture, pinpointing particularly stubborn problems and proposing solutions to enhancing our flow of knowledge. We face a deluge of information that is not always sound, accurate, or contextualized, and these sound bites can go viral in a snap of the fingers. What's more, we have seen a growing polarization that pits different groups of people against one another, creating an "us versus them" mentality that inhibits communication.

Communication is a powerful tool, one we can use well or poorly. It can yield tremendous progress, and, if we make the effort to both speak and listen carefully about the challenges outlined in this book, we will be well on our way to solving them. However, if we don't invest wisely in our communication, the way we talk about food will not only fail us, it will also fail those who have no voice at all. And these people, as the next chapter illustrated, face dire challenges indeed.

In the book's final chapter, chapter 12, Gerald Shively tackled what we may see as, really, the ultimate challenge that integrates all that came before—achieving equal access to food. Some people spend their days worried about where and how they will get their next meal. Still others obsess over posting their perfectly composed social media photos of beautifully plated, delicious food. In this world of obvious contradictions, what factors determine who gets to eat what, when, and how much? Jerry took readers on a tour of food inequality that stretched from Kathmandu to Copenhagen, and from caviar to corn, examining the contributing roles of income, prices, and public action in determining the size and composition of someone's next meal.

Ensuring sound food and nutrition for all is the task at hand. And while we face no dearth of challenges in our quest to feed the world, we also face no shortage of people who care. We don't all have to be scientists. We can also be writers, like Jessica, or perhaps even a honeybee farmer, like myself, your editor Ken. People across a wide range of skill sets and talents are committed to helping—something

I have occasion to know firsthand. Alongside my honeybee business, lovingly tended on evenings and weekends, is of course my teaching, where every spring semester for 16 weeks myself and a group of undergraduate students grind out an intense agricultural economics course. Without fail, each year I confront faces eager to tackle big challenges, and they are uniquely prepared to do it. Their belief in what is possible is not jaded by disappointing experiences. They have a host of new technologies at their disposal. Like our population expert Brigitte, they are extremely optimistic, even in the face of the grand challenges presented by this book.

In their case, it is not because they won't be here to live in this future world, because they will be. It is because they really and truly believe that we are equipped with the abilities and knowledge to change the way we interact with each other and our planet. They are not daunted by the task of sustainably feeding as many as ten billion people by the year 2050 while combating climate change and global-scale inequalities. They see it as their life's work, as fundamental to their careers and future successes. Not a single one has ever shrank from this harsh reality, nor talked about going off the grid and living in a cabin in the woods. That, teamed with the tough but *surmountable* challenges presented in this book, should make us feel better—it certainly does for yours truly.

There is, perhaps, no finer calling, nor a calling that can potentially save so many lives, as working toward sustainable food security for our world. And so, the very, very last challenge then is one issued by us. Whether we be 90 years old or 9, there is space for everyone at the table who wants to contribute to a food secure future, in ways both big and small. And therefore, we challenge you to take your unique talents and attributes, whether they be the optimism and enthusiasm of youth or the wisdom and insights of age, as well as your education and your capacity to educate others, and join us as we work toward global food security, not just for us, but for everyone, and not just for now, but for the future.

AFTERWORD

Jayson Lusk

We have all seen the headlines that spark food debates. Bacon causes cancer. A soda tax passes in Philadelphia. California voters reject mandatory labeling of GMOs. Cholesterol is no longer an ingredient of concern. The US Congress repeals mandatory country of origin labeling on meat. Chicago bans foie gras.

Discussions on these topics often make for lively dinner conversation or perhaps even uncomfortable family holidays. The issues salient enough to grab the headlines, however, are not always those that will fundamentally affect our long-term well-being. In many cases, the events of the day can distract from deeper concerns.

As a case in point, over the past decade, I have been involved in numerous public discussions about the use of biotechnology (i.e., genetically modified organisms, or "GMOs") in agriculture. Although there are serious scientific issues and discussions about the technology, my experience has been that an individual's support or opposition to GMOs often has little to do with the technology itself. Rather, it is often the case that people are concerned about deeper issues, such as whether food production will keep pace with population growth, the benefits of innovation are equitably distributed across the food supply chain, or pesticides are causing undue environmental harm. Yet, because these topics are seldom discussed on their own merits, tensions and concerns bubble up and manifest in debates about whether there should be mandatory labeling for GMOs. In short, GMOs have become a lightning rod for larger debates about the future of agriculture.

What is true for GMOs holds for many other issues as well. We fight over whether a locale should enact a soda tax, but underneath the policy divide are more fundamental concerns about high rates of obesity and the growing prevalence of diabetes and ideological tensions that pit public health concerns against freedom of choice.

Beyond the attention-grabbing headlines are serious issues driving concerns about food and agriculture. The chapters in this volume represent an attempt to elevate the discussion, focusing attention on those challenges that are likely to have the biggest impact on our food and agricultural futures. More than just raising attention, the authors make a concerted effort to ask what might be done to solve the pressing issues of our time.

In 2015, I asked the following of a group of 1,000 people: *Thinking about the future, which of the following food and agriculture challenges are you most concerned about?* They had nine items to rank. The most concerning issue, ranking highest for almost a quarter of respondents, was "having affordable food for me and my family." This was followed by 12 percent of respondents who gave "changing the type and quantity of food eaten to address obesity, diabetes, and heart disease" the highest rank. Of next-highest concern were "producing enough food to meet the demands of a growing world population" and "finding ways to prevent adverse impacts of food production," estimated to be of highest concern for 11 percent and 10 percent of respondents, respectively.

As the chapters in this book illustrate, we already know a lot and have a good start on making informed choices about these and other issues. We also have the capacity to meet these challenges. Let's get to work.

ACKNOWLEDGMENTS

The writing of this book was built off the labor of, apart from your two editors, 15 individuals. Before any other thanks, a hearty acknowledgment goes out to these hardworking folks for their time and contributions.

Beyond these contributors, many people aided this effort along with providing their support. Endless thanks to Thilo Balke, Laura Eise, Johanna Wu, Louise Eise, and Aiden Powell for their support and feedback. A big nod to two committed interns, Kayla Groen and Pamela Kuechenmeister, for their eager involvement.

Gerald Shively (chapter 12) would like to acknowledge the Purdue Policy Research Institute and the Feed the Future Innovation Lab for Nutrition, both of which are funded by the United States Agency for International Development. The opinions expressed in his chapter are his alone and do not necessarily reflect the views of these organizations.

Sadly, over the time in which this book was conceptualized and completed, we lost two dear colleagues, Corinne Alexander and Raymond Florax, to untimely deaths. Both of them would have enjoyed participating in this project. Corinne's expertise on the economics of food and agricultural storage would have been a valuable contribution to the food waste and loss discussion. Raymond was passionate about the work his wife, Brigitte Waldorf, was doing on population, which is shared in the very first chapter of the book. He also had valuable expertise on land and its uses as they vary over space and economic conditions. We miss them both, and are saddened they are no longer with us.

Lastly, thanks to our publisher, Island Press, for believing in this project, and to our lovely editor, Emily Turner. And, of course, there is Purdue University's College of Agriculture. A final, giant thanks to them and the support we found therein. In particular, thanks to Jay Akridge and Maureen Manier for their belief in this project, and their willingness to go the extra mile in support of it.

NOTES

Chapter 1. Inhabitants of Earth

1. Mark Lino, et al., *Expenditures on Children by Families, 2015*, Miscellaneous Publication 1528-2015 (Washington, DC: US Department of Agriculture, Center for Nutrition Policy and Promotion, 2017).
2. *Contraceptive prevalence* refers to the percentage of women who are currently using, or whose sexual partner is currently using, at least one method of contraception, regardless of the method used. The unmet need for family planning is usually reported for married or in-union women aged 15 to 49. See the United Nations database at http://www.un.org/en/development/desa/population/publications/dataset/contraception/wcu2014.shtml.
3. United Nations, Department of Economic and Social Affairs, Population Division, *World Population Prospects: The 2017 Revision, DVD Edition* (New York, NY: United Nations, Department of Economic and Social Affairs, 2017).
4. R. Woods and C. W. Smith, "The Decline of Marital Fertility in the Late Nineteenth Century: The Case of England and Wales," *Population Studies* 37, no. 2 (1983): 207–25.
5. The World Bank's country classification distinguishes among low-income countries (e.g., Haiti), lower-middle income countries (e.g., Honduras), upper-middle income countries (e.g., Mexico), and high-income countries (e.g., Germany), https://datahelpdesk.worldbank.org/knowledgebase/articles/906519-world-bank-country-and-lending-groups.
6. World Bank, "Life Expectancy at Birth, Total (Years)," http://data.worldbank.org/indicator/SP.DYN.LE00.IN?locations = MX.
7. According to the International Data Base of the US Census Bureau Niger's total fertility rate is predicted to decline to about 3.5 babies per woman in the next 25 years. https://www.census.gov/population/international/data/idb/informationGateway.php.
8. Pauline Vidal, "The Emigration of Health-Care Workers: Malawi's Recurring Challenges," *Online Journal of the Migration Policy Institute*, http://www.migrationpolicy.org/article/emigration-health-care-workers-malawis-recurring-challenges.
9. Edward Glaeser, "Cities, Productivity, and Quality of Life," *Science* 333, no. 6042 (2011): 592–94.
10. Data on the city size distribution have been taken from the United Nations publication *The World's Cities in 2016*, http://www.un.org/en/development/desa/population/publications/pdf/urbanization/the_worlds_cities_in_2016_data_booklet.pdf.

Chapter 2. The Green, Blue, and Gray Water Rainbow

1. Jennifer Ann Roath, "An Evaluation of Spatial Variability of Water Stress Index across the United States: Implications of Supply and Demand in the East vs the West," master's thesis, Purdue University (2013).
2. "2017 Infrastructure Report Card: 'D' Dams," accessed April 23, 2017, http://www.infrastructurereportcard.org/cat-item/dams/.
3. Hipólito Medrano et al., "From Leaf to Whole-Plant Water Use Efficiency (WUE) in Complex Canopies: Limitations of Leaf WUE as a Selection Target," *Crop Journal* 3 (2015): 220–28, doi: 10.1016/j.cj.2015.04.002.
4. Tony Allan, "'Virtual Water': A Long Term Solution for Water Short Middle Eastern Economies?" Paper presented at the 1997 British Association Festival of Science, University of Leeds, September 9, 1997.
5. A. K. Chapagain and A. Y. Hoekstra, *Water Footprints of Nations*, Value of Water Research Report Series 16 (Delft, the Netherlands: UNESCO-IHE, 2004), 80.
6. Katherine Boehrer, "This Is How Much Water It Takes to Make Your Favorite Foods," last modified April 13, 2015, http://www.huffingtonpost.com/2014/10/13/food-water-footprint_n_5952862.html.
7. A. Y. Hoekstra, The Hidden Water Resource Use behind Meat and Dairy, *Animal Frontiers* 2, no. 2 (2012): 3–8.
8. Rick Rasby, "Determining How Much Forage a Beef Cow Consumes Each Day," University of Nebraska, Lincoln, 2013, http://beef.unl.edu/cattleproduction/forageconsumed-day.
9. Brian Brettschneider, "Intra-annual Climate Variability," from Brian B's Climate Blog, December 2, 2014, http://us-climate.blogspot.com/2014/12/intra-annual-climate-variability.html.
10. Daniel Hillel, *Introduction to Soil Physics* (San Diego: Academic Press, 1982).
11. Jim Yardley, "Chinese Dam Projects Criticized for Their Human Costs," *New York Times*, November 19, 2007, http://www.nytimes.com/2007/11/19/world/asia/19dam.html.
12. Cheri Gayfield, "What Happens When the Well Runs Dry?" *Newton County Enterprise*, August 14, 2012, http://www.newsbug.info/newton_county_enterprise/news/local/what-happens-when-the-well-runs-dry/article_045b209e-e622-11e1-953a-0019bb2963f4.html.
13. David Mercer, "Many Well Users Find Their Faucets Are Running Dry," Associated Press, August 14, 2012, http://usatoday30.usatoday.com/weather/drought/story/2012-08-14/drought-dry-wells/57054870/1.
14. Molly A. Maupin and Nancy L. Barber, *Estimated Withdrawals from Principal Aquifers in the United States*, U.S. Geological Survey Circular 1279 (Washington, DC: USGS, 2000), 52.
15. E. D. Gutentag et al., "Geohydrology of the High Plains Aquifer in Parts of

Colorado, Kansas, Nebraska, New Mexico, Oklahoma, South Dakota, Texas, and Wyoming," USGS Professional Paper 1400-B (Washington, DC: USGS, 1984), 63.

16. V. L. McGuire, *Water-Level Changes in the High Plains Aquifer, Predevelopment to 2007, 2005–06, and 2006–07*, U.S. Geological Survey Scientific Investigations Report 2009-5019 (Washington, DC: USGS, 2009), 18.
17. Marios Sophocleous and Dan Merriam, "The Ogallala Formation of the Great Plains in Central US and Its Containment of Life-Giving Water," *Natural Resources Research* 21 (2012): 415–26, doi: 10.1007/s11053-012-9190-4.
18. Naresh Devineni and Shama Perveen, "Securing the Future of India's 'Water, Energy and Food,'" Global Water Forum, Discussion Series, 2012, http://www.globalwaterforum.org/2012/10/08/securing-the-future-of-indias-water-energy-and-food/.
19. "Reducing Water Use in Israel," https://water.fanack.com/israel/reducing-water-use/.
20. Thomas C. Brown and Pamela Froemke, "Nationwide Assessment of Non-point Source Threats to Water Quality," *BioScience* 62 (2012): 136–46.
21. National Research Council, *Mississippi River Water Quality and the Clean Water Act: Progress, Challenges, and Opportunities* (Washington, DC: National Academies Press, 2008).
22. "Hypoxia in the Northern Gulf of Mexico," http://www.gulfhypoxia.net.
23. R. J. Diaz and R. Rosenberg, "Spreading Dead Zones and Consequences for Marine Ecosystems," *Science* 321 (2008): 926–29, doi: 10.1126/science.1156401.
24. J. Watts, "Mexico City's Water Crisis—from Source to Sewer," *Guardian*, November 12, 2015, https://www.theguardian.com/cities/2015/nov/12/mexico-city-water-crisis-source-sewer.

Chapter 3. The Land That Shapes and Sustains Us

1. Michael Bean, Robert Bonnie, Tim Male, and Tim Searchinger, *The Private Lands Opportunity: The Case for Conservation Incentives* (New York, NY: Environmental Defense, 2003), 21.
2. Rattan Lal, Ronald Follett, and J. M. Kimble, "Achieving Soil Carbon Sequestration in the US: A Challenge to Policy Makers," *Soil Science* 168 (2003): 827–45.
3. J. D. Garbrecht et al., "Impact of Weather and Climate Scenarios on Conservation Assessment Outcomes," *Journal of Soil and Water Conservation* 69 (2014): 374–92.
4. Ralph E. Heimlich and William D. Anderson, *Development at the Urban Fringe and Beyond: Impacts on Agriculture and Rural Land*, Agricultural Economic Report 803 (Washington, DC: US Department of Agriculture, 2001), 88.

5. Natural Resources Conservation Service, Washington, DC, and Center for Survey Statistics and Methodology, Iowa State University, *Summary Report: 2012 National Resources Inventory* (Washington, DC: US Department of Agriculture, 2015), https://www.nrcs.usda.gov/Internet/FSE_DOCUMENTS/nrcseprd396218.pdf.
6. US Department of Agriculture and President's Council on Environmental Quality, *National Agricultural Lands Study* (Washington, DC: US Government Printing Office, January 1981).
7. See www.farmlandinfo.org for more ideas and tools that help protect farmland.

CHAPTER 4. OUR CHANGING CLIMATE

1. Jean Jouzel et al., "Orbital and Millennial Antarctic Climate Variability over the Past 800,000 Years," *Science* 317 (2007): 793–96, doi: 10.1126/science.1141038.
2. Intergovernmental Panel on Climate Change, *Climate Change 2013: The Physical Science Basis*, Contribution of Working Group I to the Fifth Assessment Report of the Intergovernmental Panel on Climate Change (New York: Cambridge University Press, 2013).
3. John Walsh et al., "Our Changing Climate," in *Climate Change Impacts in the United States: The Third National Climate Assessment* (Washington, DC: U.S. Global Change Research Program, 2014), 19–67, doi: 10.7930/J0KW5CXT.
4. Jerry Hatfield et al., "Climate Impacts on Agriculture: Implications for Crop Production," *Agronomy Journal* 103 (2011): 351–70, doi: 10.2134/agronj2010.0303.
5. Irakli Loladze, "Rising Atmospheric CO_2 and Human Nutrition: Toward Globally Imbalanced Plant Stoichiometry?" *Trends in Ecology & Evolution* 17 (2002): 457–61, doi: 10.1016/S0169-5347(02)02587-9.
6. Daniel R. Taub, Brian Miller, and Holly Allen, "Effects of Elevated CO_2 on the Protein Concentration of Food Crops: A Meta-Analysis," *Global Change Biology* 14 (2008): 565–75, doi: 10.1111/j.1365-2486.2007.01511.x.
7. Theodore H. Tulchinsky, "Micronutrient Deficiency Conditions: Global Health Issues," *Public Health Reviews* 32 (2010): 243–55.
8. Intergovernmental Panel on Climate Change, *Climate Change 2014: Impacts, Adaptation, and Vulnerability*, Contribution of Working Group II to the Fifth Assessment Report of the Intergovernmental Panel on Climate Change (New York: Cambridge University Press, 2014).
9. David B. Lobell et al., "Greater Sensitivity to Drought Accompanies Maize Yield Increase in the U.S. Midwest," *Science* 344 (2014): 516–19, doi: 10.1126/science.1251423.
10. Nienke Beintema et al., *ASTI Global Assessment of Agricultural R&D Spending:*

Developing Countries Accelerate Investment, International Food Policy Report (Washington, DC: International Food Policy Research Institute, Agricultural Science and Technology Indicators, Global Forum on Agricultural Research, 2012).

11. Tamma A. Carleton and Solomon M. Hsiang, "Social and Economic Impacts of Climate," *Science* 353 (2016), doi10.1126/science.aad9837.
12. Robert Kopp, Jonathan Buzan, and Matthew Huber, "The Deadly Combination of Heat and Humidity," *New York Times*, June 6, 2015, https://www.nytimes.com/2015/06/07/opinion/sunday/the-deadly-combination-of-heat-and-humidity.html?_r=0.
13. Yan Zhao et al., "Potential Escalation of Heat-Related Working Costs with Climate and Socioeconomic Changes in China," *Proceedings of the National Academy of Sciences* 113 (2016): 4640–45, doi: 10.1073/pnas.1521828113.
14. Olli Seppänen, William J. Fisk, and Q. H. Lei, "Effect of Temperature on Task Performance in Office Environment" (Berkeley, CA: Ernest Orlando Lawrence Berkeley National Laboratory, 2006).
15. Geoffrey Heal and Jisung Park, "Temperature Stress and the Direct Impact of Climate Change: A Review of an Emerging Literature," *Review of Environmental Economics and Policy* 10 (2016), doi: 10.1093/reep/rew007.
16. *Effect of Environment on Nutrient Requirements of Domestic Animals* (Washington, DC: National Academies Press, 1981).
17. John Tyndall, "On Radiation through the Earth's Atmosphere," *Journal of the Franklin Institute* 77, no. 6 (1864): 413–18.
18. C. Le Quéré et al., "Global Carbon Budget 2015," *Earth System Science Data* 7 (2015): 349–96, doi: 10.5194/essd-7-349-2015, 2015.
19. NOAA National Centers for Environmental Information, State of the Climate, "Global Analysis—Annual 2016," 2017, https://www.ncdc.noaa.gov/sotc/global/201613.
20. K. E. Taylor, R. J. Stouffer, and G. A. Meehl, 2012. "An Overview of CMIP5 and the Experiment Design," *Bulletin of the American Meteorological Society* 93, no. 4 (2012): 485–98.
21. Jonah Busch et al., "Structuring Economic Incentives to Reduce Emissions from Deforestation within Indonesia," *Proceedings of the National Academy of Sciences* 109 (2012), doi: 10.1073/pnas.1109034109.
22. Elizabeth Barona et al., "The Role of Pasture and Soybean in Deforestation of the Brazilian Amazon," *Environmental Research Letters* 5 (2010), doi: 10.1088/1748-9326/5/2/024002.
23. Daniel Nepstad et al., "Slowing Amazon Deforestation through Public Policy and Interventions in Beef and Soy Supply Chains," *Science* 34 (2014): 1118–23, doi: 10.1126/science.1248525.

24. Alla Golub et al., "The Opportunity Cost of Land Use and the Global Potential for Greenhouse Gas Mitigation in Agriculture and Forestry," *Resource and Energy Economics* 31 (2009): 299–319, doi: 10.1016/j.reseneeco.2009.04.007.
25. Alla Golub et al., "Global Climate Policy Impacts on Livestock, Land Use, Livelihoods, and Food Security," *Proceedings of the National Academy of Sciences* (2016): 1–6, doi: 10.1073/pnas.1108772109.
26. Mark Z. Jacobson and Mark A. Delucchi, "Providing All Global Energy with Wind, Water, and Solar Power, Part I: Technologies, Energy Resources, Quantities and Areas of Infrastructure, and Materials," *Energy Policy* 39 (2011): 1154–69, doi: 10.1016/j.enpol.2010.11.040.
27. William Nordhaus, *A Question of Balance: Weighing the Options on Global Warming Policies* (New Haven: Yale University Press, 2008).

Chapter 5. The Technology Ticket

1. Data from the Food and Agriculture Organization of the United Nations, 2015.
2. Rienk R. van der Ploeg, W. Böhm, and Mary Beth Kirkham, "On the Origin of the Theory of Mineral Nutrition of Plants and the Law of the Minimum," *Soil Science Society of America Journal* 63 (1999): 1055–62, doi: 10.2136/sssaj1999.6351055x.
3. Donald N. Duvick, "Biotechnology in the 1930s: The Development of Hybrid Maize," *Nature Reviews Genetics* 2 (2001): 69–74, doi: 10.1038/35047587.
4. A. L. Olmstead and P. W. Rhode, "Reshaping the Landscape: The Impact and Diffusion of the Tractor in American Agriculture, 1910–1960," *Journal of Economic History* 61 (2001): 663–98.
5. M. R. Cooper, G. T. Barton, and A. P. Brodell, *Progress of Farm Mechanization* (Washington, DC: US Department of Agriculture, 1947).
6. Food and Agriculture Organization (FAO), *The State of Food and Agriculture* (Washington, DC: FAO, 1968).
7. John H. Perkins, "The Rockefeller Foundation and the Green Revolution, 1941–1956," *Agriculture and Human Values* 7 (1990): 6–18, doi: 10.1007/BF01557305.
8. Dana G. Dalrymple, "The Development and Adoption of High-Yielding Varieties of Wheat and Rice in Developing Countries," *American Journal of Agricultural Economics* 67 (1985): 1067–73, doi: 10.2307/1241374.
9. Gurdev S. Khush, "Rice Breeding: Past, Present and Future," *Journal of Genetics* 66 (1987): 195–216, doi: 10.1007/BF02927713.
10. Chris P. Reij and E. M. A. Smaling, "Analyzing Successes in Agriculture and Land Management in Sub-Saharan Africa: Is Macro-level Gloom Obscuring Positive Micro-level Change?" *Land Use Policy* 25 (2008): 410–20, doi: 10.1016/j.landusepol.2007.10.001.

11. R. E. Evenson and D. Gollin, "Assessing the Impact of the Green Revolution, 1960 to 2000," *Science* 300 (2003): 758–62, doi: 10.1126/science.1078710.
12. Keijiro Otsuka and Donald F. Larson, "Conclusions: Strategies towards a Green Revolution in Sub-Saharan Africa," in *In Pursuit of an African Green Revolution*, ed. Keijiro Otsuka and Donald F. Larson, Natural Resource Management and Policy (New York: Springer, 2016), 183–94.
13. David A. Raitzer and Timothy G. Kelley, "Benefit–Cost Meta-Analysis of Investment in the International Agricultural Research Centers of the CGIAR," *Agricultural Systems* 96, no. 1 (2008): 108–23.
14. International Food Policy Research Institute (IFPRI), *Green Revolution: Curse or Blessing?* (Washington, DC: IFPRI, 2002), 1–4.
15. J. E. Casida and G. B. Quistad, "Golden Age of Insecticide Research: Past, Present, or Future?" *Annual Review of Entomology* 43, no. 1 (1998): 1–16, https://doi.org/10.1146/annurev.ento.43.1.1.
16. C. Milesia et al., "Mapping and Modeling the Biogeochemical Cycling of Turf Grasses in the United States," *Environmental Management* 36 (2005): 426–38, doi: 10.1007/s00267-004-0316-2.
17. H. Willer and J. Lernoud, eds., *The World of Organic Agriculture: Statistics and Emerging Trends 2015* (Frick, Switzerland: Research Institute of Organic Agriculture [FiBL], 2015), http://www.organic-world.net/yearbook/yearbook2015.html.
18. John P. Reganold and Jonathan M. Wachter, "Organic Agriculture in the Twenty-First Century," *Nature Plants* 2, no. 2 (2016): nplants2015221, doi: 10.1038/nplants.2015.221.
19. F. J. Pierce, P. Nowak, and P. C. Roberts, *Aspects of Precision Agriculture*, Advances in Agronomy 67 (New York: Academic Press, 1999).
20. The Future of Agriculture, *Technology Quarterly* June 11, 2016, http://www.economist.com/technology-quarterly/2016-06-11.
21. David Schimmelpfennig and Robert Ebel, *On the Doorstep of the Information Age: Recent Adoption of Precision Agriculture*, Economic Information Bulletin EIB-80 (Washington, DC: USDA Economic Research Service, 2011), https://www.ers.usda.gov/publications/pub-details/?pubid=44576.
22. Alexander Gogos, Katja Knauer, and Thomas D. Bucheli, "Nanomaterials in Plant Protection and Fertilization: Current State, Foreseen Applications, and Research Priorities," *Journal of Agricultural and Food Chemistry* 60 (2012): 9781–92, doi: 10.1021/jf302154y.
23. Mariya Khodakovskaya et al., "Carbon Nanotubes Are Able to Penetrate Plant Seed Coat and Dramatically Affect Seed Germination and Plant Growth," *ACS Nano* 3 (2009): 3221–27, doi: 10.1021/nn900887m.
24. M. Nuruzzaman et al., "Nanoencapsulation, Nano-guard for Pesticides: A

New Window for Safe Application," *Journal of Agricultural and Food Chemistry* 7 (2016): 1447–83, doi: 10.1021/acs.jafc.5b05214.

25. Teodoro Stadler, Micaela Buteler, and David K. Weaver, "Novel Use of Nanostructured Alumina as an Insecticide," *Pest Management Science* 66 (2010): 577–79, doi: 10.1002/ps.1915.
26. National Academies of Sciences, Engineering, and Medicine, *Genetically Engineered Crops: Experiences and Prospects* (Washington, DC: The National Academies Press, 2016).
27. Susanne von Caemmerer, W. Paul Quick, and Robert T. Furbank, "The Development of C4 Rice: Current Progress and Future Challenges," *Science* 336 (2012): 1671–72, doi: 10.1126/science.1220177.
28. Luisa Bortesi and Rainer Fischer, "The CRISPR/Cas9 System for Plant Genome Editing and Beyond," *Biotechnology Advances* 33 (2015): 41–52.
29. National Academies of Sciences, Engineering, and Medicine, *Genetically Engineered Crops: Experiences and Prospects* (Washington, DC: The National Academies Press, 2016).

Chapter 6. Systems

1. Greg Bensinger and Laura Stevens, "Amazon to Expand Grocery Business with New Convenience Stores," *Wall Street Journal*, October 12, 2016, https://www.wsj.com/articles/amazon-to-expand-grocery-business-with-new-convenience-stores-1476189657.
2. From IBISWorld Reports.
3. S. C. Walpole et al., "The Weight of Nations: An Estimation of Adult Human Biomass," *BMC Public Health* 12 (2012): 439, doi: 10.1186/1471-2458-12-439.
4. Y. Zhao et al.,"Environmental Assessment of Three Egg Production Systems, Part I: Monitoring System and Indoor Air Quality," *Poultry Science* 94 (2015): 518–33, doi: https://doi.org/10.3382/ps/peu076.
5. Pew Charitable Trusts, "Issue Brief—Trends in U.S. Antibiotic Use, New Data Needed to Improve Prescribing, Combat Threat of Antibiotic Resistance," March 22, 2017, http://www.pewtrusts.org/en/research-and-analysis/issue-briefs/2017/03/trends-in-us-antibiotic-use.

Chapter 7. Tangled Trade

1. All statistics in this section are drawn from Purdue University's Global Trade Analysis Project data base (Aguiar et al., 2016).
2. Mark J. Gehlhar and Anita Regmi, "Factors Shaping Global Food Markets," in *New Directions in Global Food Markets*, Agriculture Information Bulletin 794 (Washington, DC: US Department of Agriculture, Economic Research Service, 2005).

3. George E. P. Box, "Science and Statistics," *Journal of the American Statistical Association* 71 (1976): 791–99, doi: 10.2307/2286841.
4. Uris L. C. Baldos and Thomas W. Hertel, "Looking Back to Move Forward on Model Validation: Insights from a Global Model of Agricultural Land Use," *Environmental Research Letters* 8 (2013): 034024, doi: 10.1088/1748-9326/8/3/034024.
5. Uris L. C. Baldos and Thomas W. Hertel, "Debunking the 'New Normal': Why World Food Prices Are Expected to Resume Their Long Run Downward Trend," *Global Food Security* 8 (2016): 27–38, doi: 10.1016/j.gfs.2016.03.002.
6. David Gale Johnson, *World Agriculture in Disarray* (New York: Macmillan, 1973).
7. Kym Anderson, Gordon Rausser, and Johan Swinnen, "Political Economy of Public Policies: Insights from Distortions to Agricultural and Food Markets," *Journal of Economic Literature* 51 (2013): 423–77, doi: 10.1257/jel.51.2.423.
8. Will Martin and Alan Winters, *The Uruguay Round and the Developing Countries* (Cambridge: Cambridge University Press, 1996), http://www.gtap.agecon.purdue.edu/resources/res_display.asp?RecordID=516.
9. Jason H. Grant, Everett B. Peterson, and Sharon Sydow, "Impacts of the Trans-Pacific Partnership for US and International Dairy Trade," paper presented at the annual meeting for the Agricultural and Applied Economics Association, Boston, Massachusetts, July 31–August 2, 2016.
10. Jason H. Grant, Thomas W. Hertel, and Thomas F. Rutherford, "Dairy Tariff-Quota Liberalization: Contrasting Bilateral and Most Favored Nation Reform Options," *American Journal of Agricultural Economics* 91 (2009): 673–84, doi: 10.1111/j.1467-8276.2009.01263.x.
11. Thomas L. Friedman, *The World Is Flat: A Brief History of the Twenty-First Century* (New York: Farrar, Straus and Giroux, 2005).
12. Nienke Beintema et al., "ASTI Global Assessment of Agricultural R&D Spending: Developing Countries Accelerate Investment," International Food Policy Report (Washington, DC: International Food Policy Research Institute, Agricultural Science and Technology Indicators, Global Forum on Agricultural Research, 2012).
13. Kym Anderson and Signe Nelgen, "Trade Barrier Volatility and Agricultural Price Stabilization," *World Development* 40 (2012): 36–48, doi: 10.1016/j.worlddev.2011.05.018.
14. Will Martin and Kym Anderson, "Export Restrictions and Price Insulation during Commodity Price Booms," *American Journal of Agricultural Economics* 94 (2012): 422–27, doi: 10.1093/ajae/aar105.
15. David H. Autor, David Dorn, and Gordon H. Hanson, "The China Shock: Learning from Labor Market Adjustment to Large Changes in Trade," National

Bureau of Economic Research Working Paper 21906 (2016), http://www.nber.org/papers/w21906.

Chapter 8. Spoiled, Rotten, and Left Behind

1. Jean C. Buzby, Hodan F. Wells, and Jeffrey Hyman, *The Estimated Amount, Value, and Calories of Postharvest Food Losses at Retail and Consumer Levels in the United States* (Washington, DC: US Department of Agriculture Economic Research Service, 2014).
2. US Environmental Protection Agency, "Sustainable Management of Food: Basics," 2017, https://www.epa.gov/sustainable-management-food/sustainable-management-food-basics#what.
3. Jenny Gustavsson et al., "Global Food Losses and Food Waste: Extent, Causes, and Prevention," report from the Food and Agriculture Organization of the United Nations, May 11, 2011, http://reliefweb.int/report/world/global-food-losses-and-food-waste-extent-causes-and-prevention.
4. Alisha Coleman-Jensen et al., *Household Food Security in the United States in 2015*, Economic Research Report ERR-215, September 2016, https://www.ers.usda.gov/publications/pub-details/?pubid=79760.

Chapter 9. Tipping the Scales on Health

1. "Obesity Consequences," Harvard T.H. Chan School of Public Health, 2017, https://www.hsph.harvard.edu/obesity-prevention-source/obesity-consequences/.
2. "Obesity and Overweight Fact Sheet," last modified June 2016, http://www.who.int/mediacentre/factsheets/fs311/en/.
3. If you are looking to achieve 7 percent body fat, while gaining 20 pounds of lean mass, you should consult reputable trainers and nutritionists familiar with your sport. People with special health needs should also consult doctors and nutritionists who are experienced in addressing their health issues. For these people, this rule does not apply.
4. Dagfinn Aune et al., "BMI and All Cause Mortality: Systematic Review and Non-linear Dose–Response Meta-analysis of 230 Cohort Studies with 3.74 Million Deaths among 30.3 Million Participants," *British Medical Journal* 353 no. 2156 (2016): 353.
5. It would be irresponsible for me to tell people just to lose weight and throw all other considerations aside. The real tangible benefits come from fat loss not just weight loss. Losing weight can be accomplished using many unhealthy methods, such as crash dieting or via various eating disorders—*not recommended*.
6. Charles Wheelan, *Naked Statistics: Stripping the Dread from the Data* (New York: WW Norton, 2013).
7. Stephan Guyenet, "Calorie Intake and the US Obesity Epidemic," last modified

April 9, 2014, http://wholehealthsource.blogspot.com/2014/04/calorie-intake-and-us-obesity-epidemic.html.

8. James Hamblin, "Science Compared Every Diet, and the Winner Is Real Food," *Atlantic*, March 24, 2014, https://www.theatlantic.com/health/archive/2014/03/science-compared-every-diet-and-the-winner-is-real-food/284595/.
9. Rena R. Wing and Suzanne Phelan, "Long-Term Weight Loss Maintenance," *American Journal of Clinical Nutrition* 82 no. 1 Suppl (2005): 222S.

Chapter 10. Social License to Operate

1. Various states have begun to propose and pass gestation stalls in recent years through propositions and other state-specific measures.
2. "What Is the Social License to Operate (SLO)?" last modified January 10, 2014, http://www.miningfacts.org/Communities/What-is-the-social-licence-to-operate/.
3. "Social License: What It Means for Food Safety," last modified September 6, 2016, http://www.foodsafetymagazine.com/enewsletter/social-license-what-it-means-for-food-safety/.
4. Jacqueline L. Nelsen, "Social License to Operate," *International Journal of Mining, Reclamation and Environment* 20 (2009): 161, doi: 10.1080/17480930600804182.
5. Robin Fox, "Food and Eating: An Anthropological Perspective," Social Issues Research Centre, http://www.sirc.org/publik/foxfood.pdf.
6. Melissa G. S. McKendree, Candace C. Croney, and Nicole J. O. Widmar, "Effects of Demographic Factors and Information Sources on United States Consumer Perceptions of Animal Welfare," *Journal of Animal Science* 92 (2014): 3161–73, doi: 10.2527/jas.2014-6874.
7. David B. Schwekhardt and William P. Browne, "Politics by Other Means: The Emergence of a New Politics of Food in the United States," *Review of Agricultural Economics* 23 (2001): 302–18, http://www.jstor.org/stable/1349950.
8. Clayton Cook-Mowery, Nicole J. Olynk, and Christopher A. Wolf, "Farm-Level Contracting for Production Process Attributes: An Analysis of rBST in Milk Production," *Journal of Food & Law Policy* 4 (2008): 177–208.
9. Jacqueline L. Nelsen, "Social License to Operate," *International Journal of Mining, Reclamation and Environment* 20 (2009): 161–62, doi: 10.1080/17480930600804182.

Chapter 11. The Information Hinge

1. Caroline Dmitri and Nielson Conklin, "The 20th Century Transformation of U.S. Agriculture and Farm Policy," Economics Information Bulletin 17, June 2005, https://www.ers.usda.gov/publications/pub-details/?pubid=44198.

2. "State of the News Media 2015," Pew Research Center, April 28, 2015, http://www.journalism.org/2015/04/29/state-of-the-news-media-2015/.
3. Jayson L. Lusk, Jutta Roosen, and Andrea Bieberstein, "Consumer Acceptance of New Food Technologies: Causes and Roots of Controversies," *Annual Review of Resource Economics* 6 (2014): 381–405.
4. Benedetto De Martino et al., "Frames, Biases, and Rational Decision-Making in the Human Brain," *Science* 313 (August 4, 2006): 684–87, doi: 10.1126/science.1128356.
5. Dana Dovey, "Why People Say No to GMO: Popular Psychology and Ethics, Not Science, Spur Dislike," *Medical Daily* (April 2015), http://www.medicaldaily.com/why-people-say-no-gmo-popular-psychology-and-ethics-not-science-spur-dislike-331126.
6. "Statement by the AAAS Board of Directors on Labeling of Genetically Modified Foods," American Association for the Advancement of Science, October 2012, http://www.aaas.org/news/statement-aaas-board-directors-labeling-genetically-modified-foods.
7. Stefaan Blancke et al., "Fatal Attraction: The Intuitive Appeal of GMO Opposition." *Trends in Plant Science* 20 (2015): 414–18, doi: 10.1016/j.tplants.2015.03.011.
8. C. Krupke et al., "Assessing the Value and Pest Management Window Provided by Neonicotinoid Seed Treatments for Management of Soybean Aphid (*Aphis glycines* Matsumura) in the Upper Midwestern United States. *Pest Management Science* 73, no. 10 (2017): 2184–93.
9. C. H. Krupke et al., "Planting of Neonicotinoid-Treated Maize Poses Risks for Honey Bees and Other Non-target Organisms over a Wide Area without Consistent Crop Yield Benefit." *Journal of Applied Ecology* 54 (2017): 1449–58, doi: 10.1111/1365-2664.12924.
10. Jessica Eise and Whitney Hodde, *The Communication Scarcity in Agriculture* (New York: Routledge, 2017).
11. Proposition 002, FollowTheMoney.org, 2015, http://www.followthemoney.org/entity-details?eid=10246623.
12. Russ Parsons, "Why Eggs Have Gotten More Expensive in California," *Los Angeles Times,* June 18, 2015, http://www.latimes.com/food/dailydish/la-dd-eggs-prices-expensive-california-20150617-story.html.

Chapter 12. Achieving Equal Access

1. Elizabeth L. Prado and Kathryn G. Dewey, "Nutrition and Brain Development in Early Life," *Nutrition Reviews* 72 (2014): 267–84.
2. Saul S. Morris, Bruce Cogill, and Ricardo Uauy, "Effective International Action against Undernutrition: Why Has It Proven So Difficult and What Can Be Done to Accelerate Progress?" *Lancet* 371: 608–21.

3. Amartya Sen, *Poverty and Famines: An Essay on Entitlement and Deprivation* (Oxford: Clarendon Press, 1981).
4. Nikos Alexandratos and Jelle Bruinsma, *World Agriculture Towards 2030/2050: The 2012 revision*, ESA Working Paper 12-03 (Rome: FAO, 2012).
5. Julie Beaulac, Elizabeth Kristjansson, and Steven Cummins "A Systematic Review of Food Deserts 1966–2007," *Preventing Chronic Disease: Pubic Health Research, Practice, and Policy* 6 (2009): 1–10.
6. Andrew Muhammad et al., *International Evidence of Food Consumption Patterns: An Update Using 2005 International Comparison Program Data*, USDA-ERS Technical Bulletin 1929 (Washington, DC: US Department of Agriculture, Economic Research Service, 2011). Note that, for low-income countries, where many people are farming to feed their families, the shares are computed in such a way that income includes the value of production, and consumption reflects the value of what is produced and eaten on the farm. A hypothetical family that worked only on their own farm, produced nothing but food, ate only what they produced, and sold nothing would have a food budget share of 1.0.
7. Barry M. Popkin "The Nutrition Transition and Obesity in the Developing World," *Journal of Nutrition* 131 (2001): 871S–73S.
8. My graph updates an earlier figure produced by Will Masters and his colleagues at Tufts University. See William A. Masters et al., "The Nutrition Transition and Agricultural Transformation: A Preston Curve Approach," *Agricultural Economics* 47 (2016): 97–114, doi: 10.1111/agec.12303.
9. You can find the most recent one at http://www.restaurant-relae.dk/wp-content/uploads/Sustainability-Report-2016_4.pdf.
10. World Health Organization, *European Food and Nutrition Action Plan 2015–2020* (Copenhagen: World Health Organization Regional Office for Europe, 2014).
11. You can learn more about Aqua at https://aqua.nasa.gov.
12. Gerald Shively, Celeste Sununtnasuk, and Molly Brown, "Environmental Variability and Child Growth in Nepal," *Health and Place* 35 (2015): 37–51.
13. Gerald Shively, "Infrastructure Mitigates the Sensitivity of Child Growth to Local Agriculture and Rainfall in Nepal and Uganda," *Proceedings of the National Academy of Sciences of the United States* 14 (2017), doi: 10.1073/pnas.1524482114.
14. The Famine Early Warning Systems Network was created by the United States Agency for International Development (USAID) in 1985 to help decision makers plan for humanitarian crises. Learn more at https://www.fews.net.

CONTRIBUTORS

JESSICA EISE is an author, doctoral candidate, and Ross Fellow in the Brian Lamb School of Communication at Purdue University. She is coauthor of *The Communication Scarcity in Agriculture* and *Against the Odds: A Path Forward for Rural America*. From 2014 to 2017, she served as director of communications for the Purdue University Department of Agricultural Economics. Her current research focuses on communication around pressing social issues, with a particular interest in participatory methodology.

KEN FOSTER is a professor of agricultural economics at Purdue University. He teaches agricultural price analysis and applied time series analysis. Foster served as head of the Department of Agricultural Economics from 2008 to 2017. He was the chair of the National Association of Agricultural Economics Administrators and has been recognized nationally for his teaching, graduate mentorship, and outreach activities. His current research focuses on coffee and cacao storage and the associated linkages to food quality, food waste, and farm household income.

URIS BALDOS is a research assistant professor of agricultural economics at Purdue University. He co-developed the Simplified International Model of agricultural Prices, Land use and the Environment (SIMPLE), a computational economic model of global agriculture that he extensively uses in his research. He also coauthored a textbook on global food sustainability. His most recent research interests are on the broad issues surrounding the global farm–food–environment nexus, local-global telecoupling, and computational economic modeling.

MICHAEL BOEHLJE is a distinguished professor of agricultural economics, and he is a fellow of the Agricultural and Applied

Economics Association and of the International Food and Agribusiness Management Association. He teaches in the executive development programs of the Center for Food and Agricultural Business and is involved in applied research for the Center for Commercial Agriculture. His most recent research has focused on alternative systems of coordination of the food and industrial product chain, innovation, and risk and uncertainty.

LAURA C. BOWLING is a professor of agronomy at Purdue University. Her research interests include quantifying the hydrologic and water quality impacts of agricultural drainage practices, investigation of edge-of-field and watershed-scale conservation measures to mitigate the impacts, and evaluation of water resources sustainability, with emphasis on agricultural water use.

KEITH A. CHERKAUER is an associate professor in agricultural and biological engineering at Purdue University who works to integrate field-based observations, remote sensing products, and hydrology models to address questions and concerns related to changes in agricultural management and to further our understanding of land–atmosphere interactions and the hydrologic cycle.

OTTO DOERING is a professor of agricultural economics at Purdue University. He is a public policy specialist on economic issues affecting agriculture, natural resources, the environment, and energy. He has worked as an adviser to the US Department of Agriculture on four farm bills, most recently on the conservation and working lands programs. He has been a member of the National Academies' Water Science and Technology Board, serves on the US Environmental Protection Agency's Science Advisory Board, and chairs the Science Advisory Board's Agricultural Science Committee.

JEFF DUKES is the Belcher Chair for Environmental Sustainability, and a professor of forestry and natural resources and biology at Purdue University. He is director of the Purdue Climate Change

Research Center and of the Boston-Area Climate Experiment, which characterizes ecosystem responses to gradients of climate change. He also leads the INTERFACE research coordination network, which connects experimentalists and modelers around the world to advance global environmental change research.

MICHAEL GUNDERSON is an associate professor of agricultural economics at Purdue University and is the associate director of research for Purdue's Center for Food and Agricultural Business. He has published on topics such as service quality in agribusiness input industries, agricultural land values, and agribusiness management. His most recent research has focused on understanding the factors that influence the financial success of agribusiness firms.

TOM HERTEL is a distinguished professor of agricultural economics and executive director of the Center for Global Trade Analysis. He is a fellow, and past president, of the Agricultural and Applied Economics Association. His most recent research has focused on the impacts of climate change and mitigation policies on global trade, land use, and poverty.

JAYSON LUSK is a researcher, writer, and speaker who serves as the head of the Purdue University Department of Agricultural Economics. He is the author of, among other works, *Unnaturally Delicious: How Science and Technology Are Serving up Superfoods to Save the World* and *Compassion, by the Pound: The Economics of Farm Animal Welfare*. As a food and agricultural economist who studies what we eat and why we eat it, since 2000 Jayson has published more than 190 articles in peer-reviewed scientific journals on a wide assortment of topics, ranging from the economics of animal welfare to consumer preferences for genetically modified food to the impacts of new technologies and policies on livestock and more.

RHONDA PHILLIPS is the dean of Purdue University's Honors College. As a planning and economic development specialist, her research focuses on community quality of life and well-being. She is a member of the College of Fellows, the American Institute of Certified Planners, and is past president of the International Society for Quality-of-Life Studies. Rhonda formerly served as a Senior Sustainability Scientist with the Wrigley Institute of Global Sustainability at Arizona State University.

GERALD SHIVELY is an associate department head and professor of agricultural economics at Purdue University. His research and teaching focus on the connections among poverty, natural resource management, and food security in developing regions of the world. He is nationally and internationally recognized for his contributions to our understanding of agricultural development and the environment.

ANN SORENSEN is the assistant vice president of programs and research director for American Farmland Trust, a national organization dedicated to preserving the nation's farm- and ranchlands. Dr. Sorensen currently directs a multiyear assessment of the current and future threats to America's remaining farm- and ranchlands. The resulting geospatial maps and policy analyses will provide a solid foundation for the long-term protection and conservation of the most valuable and most threatened farm- and ranchlands in the United States.

ARIANA TORRES is an assistant professor of horticulture and landscape architecture and agricultural economics at Purdue University. She has worked on projects looking at the impact of marketing choices on technology adoption for fruit and vegetable growers, the economic implications of social capital on entrepreneurship, and the role of community support on the resilience of small businesses after disasters. Her research is at the intersection of the horticulture industry and marketing decision-making processes.

BRIGITTE S. WALDORF is a professor of agricultural economics at Purdue University. She serves on the editorial boards of several scientific journals, including the *Journal in Regional Science*, *Papers in Regional Science*, *Annals of Regional Science*, and *International Science Review*. She has published extensively on the causes and consequences of international migration, spatial segregation, human capital, and regional growth.

NICOLE J. OLYNK WIDMAR is an associate professor of agricultural economics at Purdue University, focusing in the areas of production economics, farm management, and the intersection between consumer demand for food product attributes and on-farm production. Her most recent research incorporates the economic outcomes of an on-farm decision with the intricacies of the biological processes underlying the production system employed.

STEVEN Y. WU is an associate professor of agricultural economics at Purdue University. His work centers around contract theory and applied contracting issues in agriculture. He is a faculty affiliate in the John Glenn School of Public Affairs at Ohio State; a Research Fellow with the Institute for the Study of Labor (IZA) in Bonn, Germany; and a member of the Economic Design Network in the Department of Economics at the University of Melbourne.

INDEX

Page numbers followed by "f" indicate figures.

A

accessibility. *See* Equal access
adaptation, 65–66
Adirondack Forest Preserve (New York), 57
AeroFarms, 111
aeroponic systems, 110–111
affordability, 100
Africa
- equal access to food and, 196, 197f
- food loss and, 145–147
- Green Revolution and, 85–86
- international trade and, 123, 123f
- migration patterns and, 17–18
- nutrition transition and, 201
- population growth and, 14, 16

agricultural revolutions, 8, 78–79, 81. *See also* Green Revolution
agriculture
- climate change and, 61–67
- conventional/production, 103–107, 105f
- crop yields and, 28–32
- fossil fuel consumption from, 72–74
- land use and, 51–52
- pollution and, 43
- precision, 90
- rainfed, 32–37

Aleppo, Syria, 66
algae, 88
Allan, Tony, 30
Amazon Fresh, 98
animals, categorization of, 169–172. *See also* Livestock
antibiotic resistance, 108–109
apples, 32
Aqua satellite, 203–204
aquaponic systems, 110–111
aquifers, 39–42. *See also* Groundwater
Arab Spring, 195
Aragon, Alan, 163
Asia
- food waste and loss in, 137–138, 137f, 147
- Green Revolution and, 83–84, 85
- international trade and, 120, 122–123, 123f, 128
- nutrition transition and, 201
- population growth and, 13–14, 17–18

B

Bacillus thuringiensis (Bt), 91
Baldos, Uris, 121
Baributsa, Dieudonné, 145–146
"best used by" dates, 139
beverages, sugar-sweetened, 161
biofuels, 143–144
biotechnology. *See* Genetically modified organisms
Blakewell, Robert, 181
block pricing, 42–43
blue-water footprints, 31
Boserup, Ester, 23
Box, George, 120
brain development, 191
Brazil, 74–75
breeding. *See* Plant breeding; Selective breeding
Bt. *See Bacillus thuringiensis*
budget shares, 195–196, 197f

C

Calgene, 140
calorie density, 159, 161

"calories in vs. calories out" theory, 157, 158–159
Canada, 127
canning, 142
carbohydrate-insulin theory (CIT) of obesity, 157–159
carbon dioxide, 28–30, 64, 69–72
CEA. *See* Controlled-environment agriculture
cereal crops, 136–137, 137f
certification programs, 100–101
Chicago Board of Trade, 106
children, malnutrition and, 190–192, 193, 204–205
China
  dams and, 39
  heat stress and, 68
  international trade and, 117, 120, 122–123, 123f, 128, 130
  population growth and, 13–14
chronic malnutrition, 193
CIMP5. *See* Coupled Model Intercomparison Project
CIT theory. *See* Carbohydrate-insulin theory
civil conflict, 66–67, 193, 195
clickbait, 182, 184
climate, water scarcity and, 33–36
climate change
  agriculture and, 62–67
  causes of, 69–72
  costs to people and animals, 67–69
  as global issue, 75–76
  land use and, 52
  mitigating, 72–75
  overview of issue, 59–62, 210
  population growth and, 21
Colombia, 62–63
Colorado River, 38
commodities, 97, 106
communication
  contention, misinformation and, 181–184
  evolution of, 178–181
  future and, 187–188
  overview of, 176–178, 215
  solutions for improving, 184–187
community gardens, 112–113
community supported agriculture (CSA), 102–103
compaction, 35–36
competing food systems. *See* Supply chains
complexity, food systems and, 107–113
composting, 143
cones of depression, 40–41
consolidation, 97–99
contention, 181–184, 217–218
contraception, 5
controlled-environment agriculture (CEA), 110–112
Copenhagen, Denmark, 201–202
corn, 54, 57, 66, 105–106, 117–119, 129, 133
Coupled Model Intercomparison Project (CIMP5), 72
cover crops, 108
credence food attributes, 100–101, 109
crop insurance, subsidized, 66, 125
crop yields
  climate and, 64
  monitoring of, 90–91
  water scarcity and, 28–32
cropland, defined, 54
CSA. *See* Community supported agriculture
Cuomo, Andrew, 143

D
dairy products, 126–128, 173–174
dams, 27, 37–39
Darwin, Charles, 81
data. *See* Measurements and data
DDT, 87
decision-making, 177–178, 188
deforestation, 72–75
demographic dividends, 16–17
digesters (food), 143–144

diminishing marginal benefits, 151–152
discount food stores, 98
documentation, 100–101
drones, 90
drought, 34, 50–51, 129. *See also* Water scarcity
Dust Bowl, 50–51

E
Earth system models, 61
economic migrants, 18–19
economics
  of crop adjustments, 66
  of health, 151–153, 214
  of hunger and malnutrition, 193–196
  of international trade, 128–131
  of production agriculture, 107
  of social licenses to operate, 172–174
  of water pricing, 42–43
education, communication and, 187
efficiency, 150–151, 154
eggs, 141, 187–188
Ehrlich, Anne and Paul, 22–23
80-20 rule, 151–152, 162–163
emissions, climate change and, 69–72
engagement, communication and, 186–187
England, 9–11
equal access
  causes of lack of, 192–195
  future and, 201–206
  nutrition transition and, 196–201
  overview of issue, 189–192, 215–216
An Essay on the Principle of Population (Malthus), 77–78
Europe
  Green Revolution and, 84
  international trade and, 120, 122–123, 123f
  population growth and, 14–15
evaporation, nonproductive, 30
evapotranspiration, 33, 70
"Experiments in Plant Hybridization" (Mendel), 81–82

F
fake news, 182–183
family size, 5–6
Famine Early Warning System Network (FEWS NET), 205
famines, 193
FarmedHere, 111–112
farmer's markets, 58, 102–103
farmland, 54, 56–58. *See also* Land use
farms, defined, 105–106
farm-to-school programs, 103
Federal Trade Commission (FTC), 99
Fertility Day (Italy), 14
fertility rates, 5–6, 10–12, 13f, 21–22
fertilizers
  emissions and, 74
  integrated water resources management and, 43
  problems of, 87–88
  supply chains and, 97
  technology and, 80–81
FEWS NET. *See* Famine Early Warning System Network
fisheries, 43
flavor, 152–153
FLAVR SAVR tomatoes, 95, 140
flexibility, international trade and, 129, 130–131
food balance data, 198–201, 199f
food deserts, 194
food hubs, 103
food movement, 180–181
food safety, 100, 107, 133, 138
Food Safety and Inspection Service (USDA FSIS), 139–141
food shortages, 193
food systems. *See* Supply chains
food waste and loss, 101, 132–147, 177, 194, 213–214
Foote, Eunice, 69
forestland, 54
fossil fuels, 69–72
framing manipulation, 182
free ridership, 109

frozen foods, 140, 142
futures markets, 106

G
General Agreement on Tariffs and Trade (GATT), 116
genetically modified organisms (GMOs)
  communication about, 182
  controversy and, 217
  eating healthy and, 155, 163
  food waste and, 140
  international trade and, 124
  supply chains and, 95
  technology adoption and, 91–92
gestation stalls, 165–166, 173
Global Trade Analysis Project, 116
globalization, 59–60, 75–76
GMOs. *See* Genetically modified organisms
grain filling developmental phase, 64
Grant, Jason, 126–127
grants, 112–113
gray-water footprints, 31
Great Depression, 176
Great Plague (Eurasia), 8
Green Revolution, 83–86, 88–92
green-water footprints, 31
green-water resources, 28, 32–37
grocery stores, 97–98
groundwater, 26, 27, 39–42
growth hormone, recombinant bovine (rBGH), 173–174
guano, 81
Gulf of Mexico dead zone, 43, 107

H
harvest indices, 29
harvest losses, 133, 137f
health. *See also* Malnutrition
  avoiding obesity and, 155–161
  climate change and, 67–69
  conflicting guidelines for, 148–149
  economics approach to, 151–153
  food deserts and, 194–195
  importance of, 153–155
  nutrition transition and, 197–198
  overview of issue, 148–151, 164, 213–214
  strategy for, 161–164
health care benefits, 131
heat island effect, 20
heat waves, 67–68, 129
hermetic storage, 146
Hertel, Thomas, 121
Hillel, Daniel, 36–37, 40
Homestead Act of 1862, 50
honeybees, 183, 215–216
Hull, Cordell, 117
humidity, 67–68
hybridization, 81–82, 85
hydrologic cycle, 25
hydroponic systems, 110–111
hydropower, 27
hypoxia, 43, 88, 107

I
ICARDA, 66
ice ages, 60
immigration, 15, 17–19
income level
  equal access to food and, 195–196
  population growth and, 10–11, 15–16
India, 206–206
Indianapolis, Indiana, 71–72, 73
Indonesia, 73
Industrial Revolution, 80
influence, social licenses to operate and, 172–174
information sources, checking, 185
inheritance, laws of, 82
inputs, supply chains and, 97
insecticides, 87
insects, food loss and, 144–146
insulin, carbohydrates and, 157–159

integrated water resources
management, 42–45
interconnectivity, 208
interest groups, 173–174
Intergovernmental Panel on Climate
Change report, 65
International Comparison Program
(USDA), 195–196, 197f
International Maize and Wheat
Improvement Center, 85
international trade
current state of, 118–120
dairy, TPP and, 126–128
future of, 120–123, 123f
globalized economy and, 128–131
overview of issue, 115–118, 212–213
policy in crisis and, 123–126
Internet, 179–180. *See also*
Communication
iodine deficiency, 191
iron deficiency, 65, 191
irreversibility, 57
irrigation, 37–39, 40

J
Japan, 119, 127, 195
journalists, 179–180

K
Katz, David, 162–163
Kroger, 97–98
Kyoto Protocol, 76

L
labeling laws, 138, 139
land cover, measuring, 54–55
land use
challenges of, 51–54
climate change and, 72–75
evolving relationship with land and,
49–51
farmland requirements, 56–58
impacts of, 48–49
measurement, data and, 54–56
overview of issue, 46–48, 209–210
Latin America, 85–86
lawns, 88
licenses, 166. *See also* Social licenses to
operate
lifespan, increasing, 5
Limits to Growth (Meadows et al.), 78
livestock
categorization of, 169–172
climate change and, 68–69, 73–74
food waste and, 132–133
international trade of, 119
Lobell, David, 64
Local Food Directories website, 102
loss of food
causes and potential solutions for,
144–147
importance of, 147
overview of issue, 132–138, 213
Louisiana Purchase, 50
low mortality/low fertility regime,
10–12
low-carbohydrate diets, 157–159

M
macronutrients, 162–163
maize, heat- and drought-tolerant, 66
malnutrition, 190–196
Malthus, Thomas Robert, 22–23,
77–78
margins, 101
market, politics practiced by, 173–174
McCormick, Cyrus, 82–83
McDonald's, 98, 99, 165–166, 173
Mead, Lake, 38
measurements and data
eating healthy and, 156–157
land use and, 54–56
nutrition transition and, 198–201,
199f
precision agriculture and, 89–90
megacities, 20

Mendel, Gregor, 81–82
Mexico
  Green Revolution and, 85
  integrated water resources management and, 44–45
  international trade and, 118, 120
microbasins, 36–37
micronutrients, 162–163, 190, 191
migration, 17–19
mining of groundwater, 41
misinformation, 181–184
modeling, 61, 71, 120–121
Moderate Resolution Imaging Spectroradiometer (MODIS), 204–206
moderation, 152
mortality, 5–6
Murdock, Larry, 146–147

N
NAFTA. *See* North American Free Trade Agreement
nanotechnology, 91
National Farmers Market Directory, 102
natural disasters, population growth and, 21
NDVI. *See* Normalized Difference Vegetation Index
neonicotinoids, 183
Nepal, 189–190
New Zealand, 127
"no-change" population growth scenario, 9, 11
nonproductive evaporation, 30
Normalized Difference Vegetation Index (NDVI), 204–206
North America
  food waste and loss in, 137–138, 137f
  Green Revolution and, 84
  population growth and, 15
North American Free Trade Agreement (NAFTA), 117–118
nutrition, importance of, 190–192
nutrition transition, 196–201
nutritional content, 100

O
obesity
  avoiding, 155–161
  benefits and costs of eating to avoid, 155–161
  harm caused by, 149–150
  nutrition transition and, 197–198
  statistics on, 150
obesity paradox, 155
Occam's razor, 156
Oceania
  food waste and loss in, 137–138, 137f
  Green Revolution and, 84
  population growth and, 15
Ogallala aquifer, 41, 51
online grocery shopping, 98
optimism, 207–208
organic farming, 89, 97, 100–101, 109–110
"The Origin of Species" (Darwin), 81
ownership, land use and, 46–51

P
palatability, 159–160
palm oil, 73
Paris Climate Agreement of 2015, 76
parsimony, 156
pasture, defined, 54
perspective. *See* Subjectivity
pesticides, 80–81, 87, 107
Peterson, Everett, 126–127
Pew Charitable Trusts, 109
photosynthesis, 28, 64
PICS bags. *See* Purdue Improved Crop Storage
pigs, 132–133, 165–166, 173
plant breeding, 81–82, 91–92
plants, water scarcity and, 28–32
politics, social licenses to operate and, 173–174

pollution, 20–21, 43, 86–87
Popkin, Barry, 197
Population Bomb (Ehrlich), 78
population growth
  challenges of, 7, 21–23
  distribution of, 13–15, 15f
  inequality and, 15–17
  international trade and, 121–122
  migration and, 17–19
  overview of issue, 5–8, 207–208
  statistics and projections for, 8–13, 13f
  urbanization and, 19–21, 22
pork, 165–166, 173
postharvest losses, 133, 137f, 146
potatoes, 141
poultry, 108
Powell, John Wesley, 38
Powell, Lake, 38
precipitation events, 61
precision agriculture, 90
pricing
  equal access to food and, 195–196
  of water resources, 42–43
processed foods, 152–153, 159–160, 161–162
production agriculture, 103–107
Proposition 2 (California), 187–188
protectionism, 124–125
Puglisi, Christian, 202
Purdue Improved Crop Storage (PICS) bags, 146–147

Q
quotas, 126–128

R
rainfed agriculture, 32–37
rainwater harvesting, 36–37
rangeland, defined, 54
rBST. *See* Recombinant bovine somatropin
RCP. *See* Representative concentration pathways
reapers, 82–83
recharge, aquifers and, 39–41
recombinant bovine somatropin (rBST), 173–174
recycling, of drainage water, 36–37
Relæ (Copenhagen), 202
representative concentration pathways (RCP), 71
reservoirs, 38–39
resistance, 87, 107, 108–109
restaurants, 98–99
retirement benefits, 131
rGBH. *See* Growth hormone, recombinant bovine
rice, 85, 124, 129–130, 189–190, 192
rivers, damming of, 37–39
robustness, 156–157

S
safety (food), 100, 107, 133, 138
satellites, 204–206
satiety, 159, 161
scale, 96, 141–142
selective breeding, 181
sell-by dates, 139
Sen, Amartya, 193
Silverthorn Farm (Rossville, Indiana), 102–103
SIMPLE model, 120–122
smartphones, 88
smog alarms, 20–21
social licenses to operate
  food as status symbol and, 168–169
  influence and, 172–174
  overview of, 166–168, 214–215
  pros and cons of, 174–175
  roles of animals and, 169–172
social media, 183, 185. *See also* Communication
soils
  climate change and, 74
  land use and, 52, 55
  water scarcity and, 35–36
somatropin, 173–174
soybeans, 183

sprawl, 53
Sprengel, Carl, 80
status symbols, food as, 168–169
stomata, 64–65
storage of food, 140–141, 144–146
subjectivity, 167, 169–172, 217–218
subsidies, 66, 125
sugar-sweetened beverages, 161
supply chains
 complexity, costs, benefits and, 107–113
 complexity in, 94–96
 diversity of demands and, 99–103
 food loss, waste and, 133–134
 future and, 113–114
 overview of issue, 97–99, 211–212
 production agriculture and, 103–107
surface water resources, 26–27, 37–39
sustainability reports, 202
Sydow, Sharon, 126–127
Sysco, 99
systems. *See* Supply chains

T
tariff rate quotas (TRQ), 126–128
tariffs, 126–128
taste, 100
technology
 advances from, 80–83
 controlled-environment agriculture and, 112
 costs of, 66
 food loss and, 138
 food security and, 203–206
 future and, 92–93
 Green Revolution and, 83–86
 Greener Green Revolution and, 88–92
 impacts of, 178–179
 international trade and, 121
 land use and, 52–53
 overview of issue, 77–80, 210–211
 pitfalls of modern farm and, 86–88
 production agriculture and, 104
temperature regulation, 67–69
Tennessee Valley, flooding of, 38–39
texture, 100
tillage, 35–36
tomatoes, 94–96, 100–101, 140
TPP. *See* Trans-Pacific Partnership
tractors, 83, 90
trade. *See* International trade
traditional food attributes, 100
Trans-Atlantic Trade and Investment Partnership (TTIP), 124
Trans-Pacific Partnership (TPP), 124, 126–128
transpiration, 28–30, 33
transportation corridors, 58
TRQs. *See* Tariff rate quotas
TTIP. *See* Trans-Atlantic Trade and Investment Partnership
Tyndall, John, 69, 70
Tyson, Neil deGrasse, 33

U
uncertain futures, 57
urban farming, 112
urban heat island effect, 20
urbanization, 19–21, 22, 53–54, 179
Uruguay Round of trade talks, 115–116
US Department of Agriculture (USDA), 102, 112–113
US Foods, 99
use-by dates, 139

V
vertical farming projects, 111–112
virtual water, 30–31
von Liebig, Justus, 80

W
Wagstaff, Jesse, 107
Waldorf, Brigitte, 207–208
Walmart, 97–98, 101, 110
waste of food
 causes and potential solutions for, 138

Great Depression and, 176–177
importance of, 147
overview of issue, 132–138, 213
statistics on, 135–136
water footprints, 30–31
water resources, defined, 25–26
water resources engineers, 27–28
water scarcity
controlled-environment agriculture and, 111
crop yields and, 28–32
damming rivers and, 37–39
green-water resources and, 32–37
groundwater and, 39–42
integrated water resources management and, 42–45
overview of issue, 24–28, 26f, 208–209
water scarcity index (WSI), 26f
water use efficiencies, 29
weather, 33–36
wells, 39–42
Wheelan, Charles, 156
wildlife, 39
World Trade Organization (WTO), 116–117, 124
World War II, 84, 124–125
WSI. *See* Water scarcity index
WTO. *See* World Trade Organization

Y
Yangtze River, 39
Yaphank food digester, 143–144
yields. *See* Crop yields

Z
zinc deficiency, 65, 191

Island Press | Board of Directors